GENDER AND SPACE

GENDER AND SPACE

Femininity, Sexualization and the Female Body

Seemanthini Niranjana

Sage Publications
New Delhi ♦ Thousand Oaks ♦ London

First published in 2001 by

Sage Publications India Pvt Ltd
M-32 Market, Greater Kailash, Part-I
New Delhi 110 048

Sage Publications Inc
2455 Teller Road
Thousand Oaks, California 91320

Sage Publications Ltd
6 Bonhill Street
London EC2A 4PU

Published by Tejeshwar Singh for Sage Publications India Pvt Ltd, phototypeset by Asian Telelinks, New Delhi, and printed at Chaman Enterprises, Delhi.

Library of Congress Cataloging-in-Publication Data

Simantini.
    Gender and space: femininity, sexualization, and the female body/ Seemanthini Niranjana.
        p. cm.
    Includes bibliographical references and index.
        1. Women—Identity. 2. Femininity. 3. Sex role. 4. Body, Human. I. Title
HQ1233.S548   305.4—dc21   2001   00–051558

ISBN: 0–7619–9512–9 (US HB)
            81–7036–987–8 (India HB)

Text in 10/12 Times New Roman

Sage Production Team: Sumitra Srinivasan, Sana Aiyar, M.S.V. Namboodiri and Santosh Rawat

To the memory of my parents,
Niranjana–Anupama

Contents

Note on Transliteration

All expressions in the vernacular Kannada have been transliterated into English according to their mode of pronunciation (that is, phonetically). Diacritical marks, however, have been avoided. The meaning of local terms and phrases are rendered in the text itself where first mentioned/used.

Acknowledgements

This book has taken years to come into its own. Much like a shadow it has attached itself to me, refusing to be caught, changing shape over the years as I negotiated different turns. The ethnography on which it draws formed part of my doctoral dissertation. I thank Prof. C. Rajagopalan for his encouragement and for urging me to work on the manuscript of this book. I am also grateful to Prof. K.M. Marulasiddaiah for his contacts in the field, and to Rajendra Kumar and Jayamma for introducing me to the village of my stay. Several people in the field have breathed life into the study. I would like to thank, in particular, Honnamma, Gangakka, Revamma *ajji* and Sharadamma for their warmth and sensitivity, Kamala and Munilakshmi for their company, and the people of Aaladapalya for their large heartedness.

A number of the ideas expressed here have been raised by me earlier at various junctures: through writings, at seminars, at workshops and even the classroom. The feedback on such occasions has been very helpful, compelling me to clarify my arguments further.

I would also like to gratefully acknowledge: Tejaswini, for gentle prodding about the book and for all sorts of practical advice, but above all for moments and memories shared. My parents, who in their own way created a space for me; they would have been the happiest to see this book in print and I dedicate it to their memory. Sasheej, for being an unfailing source of strength. He has been an inseparable part of this endeavour, having lived through every moment of its writing. I owe a great deal to his critical suggestions and patience in reading through various drafts, often at the cost of his own commitments; without him, this book would never have been. And our little daughter, Ila Ananya, who has grown with this book. In more ways than I can possibly express, she has re-affirmed for me the simple but deep significance of everyday spaces.

Let us space. The art of this text
is the air it causes to circulate
between its screens.
—Glas

Introduction

To translate ethnography into writing is not just to represent it in language, but also to graft it onto another discursive terrain altogether. It can never be a simple operation. One needs to develop lines of argument that can effectively tack the two together, in the process re-arranging the ethnographic record and, what is more, re-casting the problem itself.

In Space, Across Gender

The present work is an attempt to do precisely this. It draws on my doctoral dissertation of some years ago, but my purpose here is to come to a better understanding of the spatial axis underlying everyday practices, as well as societal or group reproduction. More specifically, it is to show how considerations of spatiality can inform the *bodily* practices of women within diverse contexts and settings. Conceived in these terms, the essays comprising this book explore the conjunction of space and gender in the practices and discourses of femininity and sexuality—what I thematize as sexualization—and, in the course of affirming this theme, strive to disclose the presence of a distinctive spatial register. While perceptions of the female body and injunctions regarding female morality are central to the negotiation of this register—both within and between communities—their implications for a theorization of gender and sociality have hardly been examined. By focusing on the 'acts' within which women define their lives and the arenas, events or qualities that mark bodies as female, the primary task of this book is to offer a conceptual elaboration of how gendered bodies and spaces are produced.

It must be said, however, that the book is a product of a number of hesitations, reluctances and determinations. The reluctance of adopting a strictly monographic mode of narration that would offer a complete representation of the field; the imperative not to replicate or lapse into a thesis format, but to work the material into a readable book; and the determination to offer a conceptual rendering of, or retrieval from, the field that would bear on some cognitive frames undergirding Indian sociology/anthropology. These underlying impulses not only dictated the organization of the book but also influenced the kind of use I make of field data in my narrative. While I do draw on details obtaining from fieldwork in a village community, I do not foreground detail in a way that defines *the* ground reality. Rather, I use the former in order to draw out threads of interpretation, as modes of grounding certain conceptualizations in them, through them: *in* space, *across* gender. The attempt is to make clear—in what is said and in the manner of saying it—the ways in which femininity/sexualization assumes the translation of spatial limits and boundaries.

The fieldwork (done in 1988–89) on which this book is based was conducted in a village cluster that I will collectively call Aaladapalya, located near Bangalore. The rationale behind that ethnography (intended for a Ph.D. thesis) was to address, broadly, the issues of femininity and gender, and their social construction. It involved two aspects: one, a descriptive account of the socio-cultural constitution of gendered identities, as revealed through definitions of the feminine in a local milieu; and two, an analytical deployment of the discourse of femininity, by exploring the everyday organization of social relations and social practices in the community. The intellectual context informing that study was one in which the notion of 'gender' had emerged as a powerful revisionary one, responding to the marginality of women in prevalent analytical frameworks. The concept of gender was also based on the recognition that the category 'woman' (with its connotations of biology) could no longer serve as a starting point for analyzing the socio-historically varied processes that produced women. Theoretical trends were in the process of suggesting that the category of gender implied not merely sexual or biological differences between the male and female, but also the social recreation or construction of gendered identities. These were seen as produced by a range of socio-cultural and material practices, institutional–ideological discourses, and the practices of daily life. It was against this background that I had sought to focus on how gendered feminine subjects are formed in the context of a rural community.

As fieldwork progressed, some issues acquired clarity, but other troubling questions arose. What did become clear was the need to work towards a redefinition of the concept of gender itself. This was not a suggestion to replace gender with a more resonant indigenous category, but rather to demonstrate the complexities within the very concept of gender. It entailed, first of all, extricating oneself from pre-given conceptualizations of gender premised on inequality and sexual difference. Posing the question from a distinctly non-western cultural context brought the realization that the conjoining of gender with (in)equality resulted in the evasion of central questions of difference. It led me to ask whether gender could be more fittingly rendered as a *relational* category rather than a purely oppositional or unequal one. These issues, which arose from within the ethnography itself, were tackled at some length in the dissertation. There were other unresolved questions, however, that underlay the ethnography—primarily the question of the body and the need to have a socio-spatial perspective on it. It is these issues that I seek to centralize in this book.

To reiterate, the book's basic claim is to map the spatializing of bodily discourses and practices. Existing accounts of femininity seem inadequately equipped to undertake this task. A significant number of them either highlight socialization as central to the process of acquiring female roles, or dwell on the intricacies of the cultural construction of gender, implicitly banishing, as a result, questions of the body from their analytical frameworks. Having grappled with the limits of these terrains in my dissertation, I here seek to suggest that a *body–space* orientation should inform our studies of femininity and sexualization more strongly. At first felt only intuitively, it gradually became possible to negotiate such an orientation theoretically, while re-engaging with ethnographic materials collected from the field.

At the basis of this re-engagement lay the realization that a very strong spatial narrative governed the lives of people in the village cluster. More pertinently, much of what was said of femininity, sexualization and the female body, as well as the activities of women, was all embodied in this idiom. This recognition of the centrality of space and its constant negotiation through words and deeds provided an invaluable axis along which to speak about processes within the community. Equally, I came to be provoked by the thought that space and gender, as theorizable objects, seemed to be constituted (albeit independently of each other) by similar moves of abstraction. Space, for instance, is spoken of as that which is open, or lacking solidity, immediately evoking a strong association with emptiness. This manouevre suggests that space acquires meaning not in (and of) itself, but only in oppositional relation to adjoining or constituting entities. A strikingly similar disposition embodies gender. This is especially the case when the latter is conceptualized as a quality or identity deriving its significance from a delineation against its opposite, as in the observation that to be female is to be non-male. The important point here is that this is not a *relational* definition where one is defined in terms of the other, but an *oppositional* one, where one is defined against another. The implications of this definitional move, which confines and binds its elements within an oppositional relation, do not end here. It is also marked by a process of evaluation, not always explicit, that privileges one arm of the oppositional axis against the other. Across a wide range of discourses and disciplines one finds several such dichotomies in operation—as in mind/body, reason/emotion, culture/nature, public/private, male/female—where, invariably, the former axis is viewed as superior and creative in relation to the latter. One can add to this list a dichotomy between temporality and spatiality as well, where, again, time is perceived as more active and transformative in opposition to space, which is seen as something fixed. Such conceptions of space and definitions of femininity seem to share a common evaluative standpoint—in that both constitute the devalued part of the oppositional axis. What is more, both are also repeatedly relegated to the domain of the 'natural', whether it be space as that which is unchangeable, given for all times, or womanhood/femininity being seen as a natural (read biological) attribute (unlike, say, masculinity, which is primarily perceived as a cultural achievement).

There have, undoubtedly, been radical critiques of these positions in recent decades. However, while these signal a wider sociality–spatiality nexus, my focus is only on one aspect of it. As already indicated, I will show how the body, and the modes in which it inhabits space, itself comes to be deployed as a medium through which the 'female' is constituted. As I will argue throughout, the socio-spatial parameters of this sexualization, as indeed the dailiness of women's lives, hinge on a delineation of what is known in the vernacular (Kannada) as *olage–horage*. This spatial axis is not only deeply embedded in perceptual schemes but also decidedly orients bodily practices, ordering the world into one's own and other, designating acts as proper and improper, moral and immoral, and so on. It is an elaboration of these registers of spatiality, femininity and sexualization that will be the central focus here.

Organization of the Book

The foregoing section is perhaps indicative of the conceptual and ethnographic concerns at the heart of this work. Let me add a word about the space of its organization. Chapter One begins with the crucial question of method, especially what it means to write an ethnography of the everyday in contemporary times. Its backdrop is, of course, the radical disputations that have arisen (since the 1980s) regarding the very nature of representation, debates that precipitated a sort of crises across disciplines. The attempt by feminist ethnography to fashion alternative paths of fieldwork and description is also pertinent to my exploration. Building on these ideas to critique the monographic format, I suggested the need to work towards a narrativity that could enable us to speak, simultaneously, both *of* and *from* the field. Working in this direction, the text consists not of a linear narrative but several lateral ones, drawing on a simultaneity of relations. Despite the organization into chapters, there is a constant movement across them, each addressing issues of gender and space from a unique position and thereby recomposing the narrative at each juncture. Chapter Two seeks to elaborate on pre-existing theoretical perspectives on space and gender, only to find that unlike the critical thought of some geographers, socio-anthropological discourse has seldom addressed these issues. It also sketches the impasse within gender studies in India, suggesting that a focus on the body–space nexus could offer one way of transcending this limitation.

Drawing on and speaking off my ethnography, each subsequent chapter articulates the possible contours of a theory of gender and space from different angles. Through highlighting the domains and activities of women in the village, an attempt is made to draw out the spatial idiom underlying their lives. Through considerations of the household domain and beyond, the diverse work activities of women and the modes in which they negotiate space, it is shown how the olage and horage emerge as a crucial spatial form locating and classifying their lives (Chapter Three). Chapter Four focuses on the everyday lives of women, attempting to map how the female body is perceived, presented and lived. Facilitating this is a consideration of junctures like puberty, childbirth, and so on, which provide a dominant context for the fusion of issues like fertility and auspiciousness. The realm of ritual is of considerable significance in understanding the feminine—both in terms of deities and their qualities, and in terms of a delineation of 'appropriate' spaces for the expression of the sexuality and fertility of women (Chapter Five). Chapter Six seeks to formulate a ground for female agency, as articulated within a certain body–space matrix. Focusing on the field of disputes as well as women's gossip, it traces forms of female agency via discourses of morality. Engaging with the issues arising out of the above account and their implications for a sociological discourse of India is central to Chapter Seven. It is argued that developing and deploying a spatial matrix is indispensable for comprehending the parameters of social life in general and women's lives in particular. It also provides a way in which to script gender into social analysis. The final chapter is an attempt to work the olage–horage taxonomy into a conceptual framework. It reflects on the possibilities and challenges of articulating a sociology of the body.

One

Towards a New Narrativity

'The making of ethnography', James Clifford (1986: 6) has observed, 'is artisanal, tied to the worldly work of writing'. This statement emblematizes what is perhaps the current consensus about the constructed nature of anthropological accounts. My intention in this chapter is to touch upon two aspects of this 'consensus' as they come to bear upon my work: one, the problem of representation itself and two, an associated question about place, especially with respect to Indian ethnographies of the village. The ensuing account will also enable me to foreground a consideration long relegated to the 'undisclosed margins' of an ethnography, namely, what it means to inhabit a space disciplinarily, physically and culturally. Along this course I shall, of necessity, engage with some aspects of feminist ethnography. It is perhaps against this background that the contours of the present work would begin to emerge.

The Crisis in Representation

Since the 1980s, anthropology, among other disciplines, has been the site of fairly radical disputations regarding the very nature of representation, be it that of other cultures or, increasingly, one's own (Clifford and Marcus, 1986). A range of theoretical developments, both within and outside the discipline (for instance, semiotics, structuralism and deconstruction), have made it clear that cultural representation, especially the endeavour to provide an accurate reflection of reality, is not only an inherently problematic enterprise but also an impossibility. A review of these trends is not my concern here, though I am interested in examining their implications for the writing of anthropology—to see, in particular, what would happen to the ethnographic exercise and its written product; the monograph.

Ethnography and its written product have been the orthodox discursive forms for anthropology (perhaps even constitutive of the anthropological domain itself). One may concur with Marcus and Fischer's rather straightforward depiction of the process: 'ethnography is a research process in which the anthropologist closely observes, records, and engages in the daily life of another culture—an experience labelled as the fieldwork method—and then writes accounts of this culture, emphasizing descriptive detail' (Marcus and Fischer, 1986: 18). While the anthropologist's monograph is, by definition, fieldwork-based, what was further assumed—at least for several decades from the 1920s—was that the monograph is a kind of spontaneous form arising out of fieldwork. The concern, in other words, was overwhelmingly with the 'what' of a cultural reality rather than with 'how' to speak of it or represent it. The monographic format was very often localized to a distinct community, the endeavour being to present a holistic picture of that reality. It is this effort to represent the 'reality' of a whole way of life that has come to be described as 'ethnographic realism'. As Marcus and Fischer elaborate:

> (...) realist ethnographies are written to allude to a whole by means of parts or foci of analytical attention which constantly evoke a social and cultural totality. Close attention to detail and redundant demonstration that the writer shared and experienced this whole world are further aspects of realist writing... (Ibid.: 23).

Despite relying so strongly on the 'presence' of the ethnographer in the field, the standard monograph tended to erase this fact in the text, by ostensibly describing and presenting a reality that was somehow independent, pristine, holistic and complete. The most prevalent 'methods' in such research—participant observation, data collection, description—all suggested or posited an objectification of reality.

In recent times, however, an attentiveness to the constructed nature of such accounts and realities has succeeded in raising anew the question of 'cultural representation', both in the context of focusing on the 'making' of ethnographic texts and in assertions about the inherently partial or incomplete nature of all knowledge. These developments have come to demand a re-negotiation, not only of the task of representation but also of the very domain of anthropology. It could be argued, in this light, that the ethnographic monograph and its implicit assumptions regarding observation and representation (however much these have come to be qualified in recent writings) belong to a distinct anthropological moment. If it may indeed be admitted, as has been suggested by Manganaro

(1990: 30), that the formation of anthropology and the rise of the ethnographic monograph were mutually constitutive, then it would be instructive to go a step further to query the very form of ethnographies.

Today, when we are no longer studying societies of the past, nor alien or 'other' cultures, what happens to the classical anthropological methods? Indeed, what is the very status of the monograph in such a changed field? Writing ethnographies in the present, at a time when disciplinary domains are themselves in crises and flux, will require 'experiments with form'. As Das (1996) indicates, it is necessary to go beyond holistic studies of communities (or even efforts to record a 'vanishing tradition'), and turn instead to a deeper engagement with modernity and the question of how to write ethnographies of contemporary societies. Das's strategy in negotiating this task is to fix on certain 'critical events' that mark the contemporary scene, events that either disrupt the social fabric or which institute new modalities of action on the part of social agents—be they castes, religious groups or nations. My own focus, in contrast, is on the logic underlying everyday 'lived' worlds. Without having the high visibility of critical events, such a mundane daily logic does inform social and material practices to a significant extent. Tracing the ways in which such a logic is articulated, negotiated and/or reconstituted, and in a manner that does not replicate holistic representations of community, is a crucial task. Thus, while Das is concerned about recharting the space of the modern in specific ways, I use her observation about 'experiments with form' to argue for a changed narrative mode (elaborated in the next section). It is against this backdrop that discourses from the field are re-deployed here. This initial acknowledgement that the field offers layered economies of truth will go at least partially in facilitating a traversal of the polyvalence of everyday spaces.

Feminist Anthropology

As I indicated at the very outset, recent years have seen a fairly strong intellectual trend that has destabilized the authority of (western) anthropological representations. Feminists, among others, have focused on the implications of the historical trajectories linking the west and non-west, drawing out various strands of cultural and gender differences. The issue of location—of both the investigating subject and his/her object—has often been central to these accounts (see for instance, John, 1996; Visweswaran, 1996). Though a key issue by itself, its exploration need not engage us here. It is my concern, however, to foreground a point often implicit in these debates: that with a disruption of anthropology's founding metaphor of self/other, there also occurs a 'transformation' in the very nature of the 'field' pursued by anthropologists. No longer is this exotic, alien or far-removed, but is often the familiar, everyday world of one's own society. Such a shift has enormous implications for the practice of anthropology, both in terms of content and method. Most important is the fact that the narrative mode will also have to change, since the rationale is no longer one of 'bring(ing) an exotic place to a home audience' (Strathern, 1987b: 28). However, the precise contours such a changed mode of anthropological representation could assume have been insufficiently addressed.[1] Since feminist ethnography consti- tutes a terrain where such issues are routinely addressed, I shall turn to a consideration of this space, examining, in particular, whether it could offer some insight into the altered anthropological practices of the present.

Unlike other disciplines, say, history or even literary studies, which were targeted by feminists for their exclusion of women, the situation in anthropology has tended to be a little different. Since women have always been part of the subject-matter of anthropology, the more significant questions raised were with regard to how women had been studied, by whom, in what contexts, and so on. The problem was clearly not one of the absence of women, but of the representation of women within disciplinary frames and its adequacy or otherwise. The early feminist task was to draw attention to the muting of women within conventional anthropological discourse.[2] Various tiers of male bias that had served to relegate the study of women to certain fixed arenas and to present a male-centred view of the world were located within the discipline. The practice of fieldwork itself was identified as a major problem area, and arguments here included not only the need to focus on women subjects per se, but also of having female ethnographers conduct such studies. But as this initial flush of work in what came to be called 'anthropology of women' soon found, (Rapp [1975] is perhaps emblematic) merely generating empirical data on women in the above-mentioned fashion would not quite do; for this to be meaningful, there had to be a change in the theoretical and conceptual frameworks underlying fieldwork practice as well. It took a more self-consciously feminist anthropology (with its deployment of the gender concept) to initiate a critique of sorts by foregrounding the partial nature of truths in anthropological accounts (cf. Moore, 1988, especially Chapter One).

The discipline's response to such moves, however, has been predictable, ranging from outright marginalization to appropriation, the latter entailing an incorporation (and inevitable dilution) of certain aspects of feminist studies within dominant interpretive frames.[3] Likewise, the task of 'restoring women to view' has not turned out to be an easy one, with feminist anthropology often moving uneasily between women-centred modes of analyses and more de-centred ones, the latter focusing on women's relationality to wider societal contexts.[4]

In a bid to retain the radical edge of feminist ethnography, it has been recently argued that its potential is both theoretical and political, deriving from the move to 'locat(e) the self in the experience of oppression in order to liberate it' (Visweswaran, 1996: 19). This stance goes on to suggest that 'a feminist ethnography could focus on women's relationships to other women, and the power differentials between them' (Ibid.: 20). While such a delineation succeeds in foregrounding a number of issues, such as a focus on women, experience and power, and includes a political recognition of differences amongst women, the essentially narrative question of how to present or inscribe such ethnographies is insufficiently addressed. It seems to me that this latter question must necessarily traverse the tense space between feminism and anthropology. One may do so from within the framework of an important essay by Strathern (1987a).

Among the issues tackled by this essay, perhaps the most significant is the appraisal offered of the categories central to feminism and anthropology, namely, experience and self. Strathern points out that in spite of outwardly similar concepts being deployed by both feminism and anthropology, their effects are rather divergent. She traces this divergence to the dissimilar mobilization of self–other equations in these fields. To represent the appraisal in fairly stark terms: the field (self) of anthropology, on a conventional register, is formed in relation to an 'other', which is invariably an alien culture or society. Within experimental anthropology, however, Strathern (1987a: 289) suggests that the 'other' is not under siege, but is instead a privileged partner in the emerging dialogue/discourse; here, the representation of the 'other' becomes possible only if the self breaks with its own past, tradition and culture in the process paving the way for an erasure of the self–other distinction. Feminism, alternatively, sets up men/patriarchy as the 'other', in relation to women. As Strathern notes:

(...) creating a space for women becomes creating a space for the self, and experience becomes an instrument for knowing the self. Necessary

to the construction of the feminist self, then, is a nonfeminist other.
The other is most generally conceived as 'patriarchy', the institutions
and persons who represent male domination, often simply concretized
as 'men' (Strathern 1987a: 288).

It follows from this that feminism constitutes as well as discovers its
'self' by becoming conscious of the other's (read male) oppression.
That is to say, by the theoretico–political act of giving voice to women,
feminist scholars are rediscovering their own selves, pasts and experiences.

I must hasten to add that in drawing out the implications of Strathern's
argument, my purpose is neither to restrict feminism to the above characterization nor to deny the reflexivity exhibited by much of feminist
scholarship. Indeed, a host of issues have been engaged within these
contours—Is 'giving voice' to women to speak *for* such women? From
what positions are such ventriloquisms effected? And, how liberatory
are they? My purpose, yet, is to go along with Strathern's argument for
part of the way, to see how the fields of feminism and anthropology may
be related. If what has been discussed earlier is the self–other alignments
in/of anthropology and feminism respectively, the hybrid field of 'feminist anthropology' would be a doubly problematic one, the core question
being how it would delineate its 'other'. If it posits the 'other' as 'male',
it would begin to premise a universal sisterhood—a position that has
been viewed with great scepticism, if not an impossibility. If it posits the
'other' as 'alien culture/society', then what is the role of the 'feminist
self' defined in relation to a 'non-feminist other'? And again: who precisely are these 'non-feminist others'?

A number of strands merge here and demand further clarification. Let
me begin to disentangle the knots, while also seeking to re-entwine the
argument around a different axis. Visweswaran's delineation of feminist
ethnography in terms of the 'power differentials' between women seems
to bypass, in real terms, the tension identified by Strathern. Without
addressing the differences in self–other representations in feminism and
anthropology (or, again, the diverse ways in which the mediation of
experience could be sought), Visweswaran seems content to observe
that feminist anthropology would stand to gain if, taking a cue from
experimental anthropology, it reassessed its assumptions about the
'other'. The question as to how this would be possible without a fundamental change in the very conceptual terrain of feminist anthropology is
nowhere dwelt upon. It can, of course, be asserted that a similar silence

is discernible in Strathern's essay too, for despite posing the problem in no uncertain terms, she does not push the analysis far enough to quiz the very premises of feminist anthropology. She merely observes that anthropological studies of women either see sexual inequality as universal or adopt a more relativistic position.[5]

To my mind, it would be fruitful to engage with how the very category of feminism has become a highly unstable one in the modern context, while also implicating the terrain of anthropology more fully within this context. Assumptions of a universal (male) oppression and a universal sisterhood have been called into question by contemporary studies highlighting the differences—of race, nation, caste and class—amongst women themselves. Nevertheless, it would be instructive to graft Strathern's account of feminism not just onto the anthropological field but also into the current discourses on difference. Thus, Strathern's (1987a: 288) observation that 'necessary to the construction of the feminist self...is a nonfeminist other' when applied to the anthropological practice of a First World female ethnographer, would entail a difficulty about specifying this 'non-feminist other': Who are these 'other' women? Women in the Third World? Also, what are the implications—political and theoretical—of such a scholarship?

Clearly, then, there can be no simple answers to what an anthropologically-grounded feminist scholarship presents as the question of difference, especially since the self/other delineations get refracted at every level or layer of difference. Given these refractions, anthropology's attempt to holistically represent communities might well flounder. The main lesson that feminist ethnography affords is that a recognition of difference (of gender, culture, or whatever), must necessarily queer the pitch of representation; it must necessarily inaugurate a change in the mode of writing ethnographies as well. All this can become even more pressing and complicated when the 'field' is one's own society, an axis that would require an altered mode of grasping the self–other dichotomy.[6] Indeed, inhabiting such an altered terrain, while attempting to map ways of living in the present, would necessitate experiments both with narration and with concepts.

'Place' in Indian Village Studies

The problem of representation in anthropology that we have been raising till now is, in a sense, inseparable from the issue of location. In its most general formulation, the issue calls attention to the location of the observer vis-à-vis the society being studied. Nevertheless, despite the seeming transparency of the matter, the discursive gloss it has been subject to has varied. At times, it has merely dramatized the anthropological project of studying the 'other'. By demanding clarifications on how the observer is positioned in relation to the group being studied, it has served to critique the notion of detached observation, the 'view from nowhere' of conventional ethnography. At other times, anthropologists have attempted to run together questions of location with those of 'voice', highlighting thereby a plurality of perspectives emerging from the field and the consequent problems of representation. In recent times, problems of location have come to be raised in the company of related questions about power/ knowledge and the constitution of identities.

It could be argued, independently of these trends, that a focus on 'place' (as distinct from the politics of location) constitutes an equally significant point of departure for anthropology. Most typically, and despite its wide-ranging ruminations about cultural differences, the discipline has equated place with setting—an ethnographic locale 'where people do things'. Place also sometimes stands in, or becomes a metaphor for, certain ideas—as, for instance, the association of India with the notion of hierarchy (Appadurai, 1988b). Approached thus, place signifies 'the problem of the culturally defined locations to which ethnographies refer' (Appadurai, 1988a: 16), wherein it is almost as if place is an anthropological creation. Building a slightly different argument, a recent volume (Gupta and Ferguson, 1997a) scrutinizes both the 'field' as an anthropological site as well as the practice and methods of fieldwork. Seeking to query the 'locales' of anthropological research, it is asked whether fieldwork at home or in shifting, fluid sites, is still anthropological. In effect, these questions do transcend the dominant tendency to view the field as a passive context of/for action. But it should be mentioned that the terms of such a problematization embody a preoccupation with the very practice of anthropology. While this is without doubt an important task, such a traversal need not exhaust the meanings of place. Indeed, as a complement to the disciplinary construction of place, one would also have to address aspects of its social construction. Thus:

Places have multiple meanings (for their inhabitants) that are constructed spatially...(and) need to be understood apart from their creation as the locales of ethnography.... (More crucial is to) raise questions about how the anthropological study of place relates to experiences of living in places (Rodman, 1992: 641).

The fact that the emphasis has all along been on the politics of representation and location rather than on place per se is even illustrative of the dominant note struck in the very modes of conceptualizing space (cf. broadly Chapter Two herein). Place, with its apparent givenness, comes to be associated with stasis, with seemingly unchanging qualities, while location is seen to embody a more dynamic, fluid scene, a field of activity, a 'politics'. What is being lost sight of in this perception is the possibility that 'place' too is constructed and, what is more, can be constitutive of space. This by itself is a vast theme and can be approached from different angles. My concern, in the considerations to follow, is particularly with the epistemological dimensions thrown up by the question of place as designated by (or expressed within) Indian ethnographies of the village. This may, in turn, clarify my own incursions into the village cluster studied, while also offering an alternative evocation of that context in and through writing.

Within traditional Indian ethnographies of the village, the question of place has scarcely been more than a literal problem. Whether it be a strictly anthropological study of Shamirpet (Dube, 1967) or a quasi-fictional anthropological narration of life in Rampura (Srinivas, 1976), such accounts sought to present a holistic picture of the village, full of details about social life and social institutions therein. Indeed, though the genre of village studies played a key role in configuring the domain of sociology in India around the 1950s, sustained reflection on adopting the 'village community' as the sociological unit of study was hardly forthcoming. The locus of study, namely, the village, was largely taken for granted, and the implications of conceptualizing the latter as either a territorial, material or cultural entity remained unexamined. A scrutiny of the tradition that produced the 'village' in Indian sociology would therefore be useful as a prelude to exploring alternative ways of speaking of/from the village.

A historical perspective could well serve to ground ideas about the village while also throwing light on the power configurations within which specific perspectives on the village emerged. Indeed, the delineation of a notion like 'village community' cannot be dislodged from the larger context that defined relations between India and the west. This point could do with some elaboration. Most nineteenth century writings on the Indian village community, produced by colonial administrators, historians and political thinkers, tended to paint an idealized picture of the social, economic and political aspects of the village.[7] The standard, oft-repeated description of the village community as a 'little republic' served to highlight its self-sufficiency and isolation from the larger socio-political system. It came to be attributed with characteristics such as cohesion, political autonomy, detachment from external political currents, economic autarchy, changelessness and social homogeneity. To cite Metcalfe: 'The village communities are little republics having nearly everything that they want within themselves, and almost independent of any foreign relations...' (cited in Dumont, 1966: 73). A number of these writings were generated from, or based on, reports concerning land ownership, patterns of cultivation and other aspects. Both Marx and Maine considered the village as a necessary and early form of civilization in an evolutionary scheme of civilizational growth and progress—in short, as the repository of primitive communism. The absence of private owner-ship of land was seen as crucial to this definition of the Indian village.

The fact that the main source of knowledge about the village com-munity as it existed in the past came from British records—especially parliamentary papers, other administrative reports and British histor-ians—cannot be ignored. One cannot also fail to perceive here overtones of the Orientalist project (Said, 1985), whereby the west came to construct the east in specific cultural, ideological and material terms, in the process defining itself and the relations between the Occident and the Orient. Descriptions of the 'village community' as an autonomous, homogeneous whole were thus a product of colonial rule, necessitated by the admin-istrative needs of the British government, where smaller territorial units were crucial for easier tax collection and implementation of various policies. A recognition of this discursive and institutional context auto-matically raises the question of how one could, in the present, unproble-matically approach the village as an empirically given unit. Within a contemporary sociology of India, what could a focus on the village imply?

The bulk of village studies in India in the post-Independence period have been preoccupied with documenting, explaining and directing social transformation. Even those studies which sought to give a comprehensive account of all areas of village life worked within the reigning paradigm of social change (Dube, 1967). Their rationale ranged from the need to '...assess and evaluate the human implications (of change)' (Ibid.: 12), to that of providing 'the requisite background data from which more purposeful planning can emerge' (Dube, 1969: 793). The close nexus between such studies and the state was encapsulated in the self-assigned task of the rural sociologist, that is, to discover the laws of development of rural society so as to facilitate 'reconstruction on a more advanced economic, social and cultural basis...' (Desai, 1969: 96). Overtly or otherwise, study of the changing social situation in India was linked to developmental policies and social engineering, with a stress on planning and institutional reform. Against this general background, emphasis in some studies came to be on the unity of the village and the persistence of the social structure in spite of internal changes (Srinivas, 1955). In others, the focus was not so much on maintaining interactions within the system, but on accounting for external factors responsible for the break-up of traditional social arrangements (Marriott, 1961). Many accounts showed a deep concern with the connections between material organization, especially changes in the economy, and socio-cultural aspects (Epstein, 1979; Bailey, 1957). Studying the changes initiated in the village by larger economic and political forces also, at times, shifted the focus from the study of a single village to its extra-local links. In such efforts, the village as the sole unit of study came to be replaced by a focus on institutions and processes which merely find a 'field of operation' in the village (Madan, 1965). With the growing tide of reactions against structural functionalism, village studies also began to examine the conflictual dimensions of social life (Beteille, 1965).

Now, it should be interesting to note that across a wide range of such studies one encounters an unquestioned acceptance of the 'village' as an objective reality, a stance which has not failed to dictate the methodology guiding the analyses as well. In neither the colonialist phase with its stress on the homogeneity and self-sufficiency of villages nor in majority of the post-Independence versions of village studies' tradition do we find an in-depth reflection on the very space of the village, or the village as place per se. However, a certain engagement is forthcoming in the writings of Dumont/Pocock and Srinivas in the course of debating the epistemological grounds of undertaking sociological studies of the village. By calling attention to the village as a reference point, they contend with the basic question of what constitutes the 'sociological reality' in India. Since pursuing the ramifications of the latter question is not my intent here, I shall confine myself to those aspects of this exchange which impinge on an understanding of place in Indian ethnographies of the village.

The avowed purpose behind studying specific village communities, quite often, has been to generate socio-economic, political and religious knowledge about India (Srinivas, 1962). Consequently, villages have been approached as 'observation centres', in the context of which social processes occurring all over the country were studied. In Srinivas's view:

(...) the study of the village...provides a strategic point of entry for the study of Indian society and culture as a whole. It forces the...scholar to keep his mind...on the existential reality as contrasted with the book view of society (Srinivas, 1966: 158).

An intensive micro-study of a single village, according to him, not only produces knowledge about the textures of village life, but also offers significant insights into Indian social life as a whole. Indeed, the fashioning of tools and concepts for analyzing 'Indian culture' is also made possible by such in-depth village studies. Implied in this viewpoint is the assumption of a distinctive empirical group—in this instance, the village—as a legitimate object of study encountered through fieldwork.

The position of Dumont and Pocock differs from this one. Basically sceptical about adopting the village as a frame of enquiry, they ask whether 'the village (is) indeed the social fact which it has for so long been assumed to be?' (Dumont and Pocock, 1957: 23). The village, for them, is nothing more than a territorial and demographic fact, in itself insufficient to confer 'sociological reality' on the village; rather, they suggest that it is the social scientist, by approaching and representing the village in a specific way, who confers on it a sociological significance. According to Dumont and Pocock, there is

(...) a confusion between the necessary delimitation of a field of enquiry with the results of that enquiry. A fieldworker takes a village as a convenient centre for his investigations...(which confers) upon that village a kind of sociological reality which in fact it does not possess (Ibid.: 26).

For Dumont especially, as is well known, the route to an understanding of Indian society as a whole is through its system of ideas and common values. This concern with 'ideas and values' (that is, cultural meanings) prompts a privileging of ideas of caste and kinship as central to an understanding of Indian social organization, rather than territory per se (that is, the village). Indeed, the latter is seen as secondary in relation to the former. Thus, the village assumes importance to the extent that it is the site where cultural ideas of hierarchy, for instance, are played out.

In critical response to Dumont's insistence on a focus on ideas/values, Srinivas foregrounds fieldwork as essential in order to grasp the dynamics and complexities of Indian society and culture. He is thereby also battling against the use of textual material in analysis which, according to him, misrepresents the complex interlinking of elements in the social system. The distortions emerging from a 'book view' of Indian society are believed to be overcome by intensive fieldwork. Here again, we do find invocations of the village as field, but as a pre-given entity which submits itself to patient and persistent observation. For Srinivas (1987), 'sociological reality' lies not in ideas or values, but in the fact of community.

It seems to me that these two perspectives are indicative of a baseline alignment of positions, and contemporary scholars have extended the argument in different ways. Yet, in terms of the above juxtaposition, it should be evident that despite the polarization of these positions, both come to use the 'village' as a point of entry for understanding Indian society and for debating what constitutes the sociological reality in India.[8] Indeed, read against the background of work foregrounding these perspectives, it is clear that the 'village' has been habitually defined more in terms of territory (as a matter of administrative convenience), or material organization (the peasant economy), or cultural homogeneity (especially in invoking caste values). None of these orientations facilitate a focus on the village-as-place—or, more precisely, on the village as a socio-spatial entity—where the latter is not a mere context, but the medium and outcome of a range of socio-spatial practices.

It could therefore be remarked that in relation to the above trends in ethnographies of the Indian village, this work is somewhat off-centre. Adopting neither of the extreme positions exclusively, the task I address is one of re-assembling the contours of the village (field) in a manner which breaks with the above tendencies. In other words, the endeavour is to provide an alternative matrix for theorizing social spaces. Drawing on the manner in which space (both social and physical) is delineated in the course of everyday life in a village cluster, I focus on an under-explored matrix of space and gender, arguing that neither of these can be retrieved along purely material or purely cultural axes. The very notion of *uru* (village), as I shall indicate in the chapter 'Spatial Registers', is seldom a strictly geographical unit, but is rather a gendered, spatial, socio-cultural matrix. I shall engage with this matrix further by indicating some modes through which a mapping of space is achieved, focusing in particular on perceptions of and injunctions surrounding female bodies and female morality. These schemes of thought and practice, we shall see, are central to the negotiation of space, specifying how identities are consolidated and lived in the course of marking the boundaries of movement and the 'limits' of women's honour. The question of place will, accordingly, no longer remain a literal or practical problem, but will have to engage with a gendered socio-spatial configuration.

Perhaps some of these remarks prepare the ground for what is to follow. This book is not a monograph, in the sense of providing a detailed, holistic picture of the field, whether 'remote' or 'closer home'. It seeks less to represent cultural worlds than to show how bodies are spaced and gendered, an endeavour that would require a dialogue between indigenous concepts and analytical ones. In the chapters that follow, I examine ideas about space (variedly expressed through words, body movements, work, or ritual), discern its operation as a significant principle ordering the everyday lives of people, and, at the same time, move a step further in grasping (analytically) the use of spatial parameters in the constitution of female bodies and female morality. If I speak as an insider, it is only as a partial insider, for I am simultaneously also poised outward—discursively, institutionally and existentially—a situation that only underscores the significance of differences within India, where the 'inside' is never fixed nor unitary, but always the product of its changing relations with the specific constellations of the 'outside'.

Notes

1. Indeed, the possibility of studying one's own society was envisaged and practiced by some early Indian social anthropologists (see Srinivas, 1966 and 1996b). Yet, in spite of having provoked ruminations about subjectivity, knowledge of the insider and preconceptions of the researcher, a full-throated critique and transformation of the representational mode in anthropology did not follow. (See also Note 6 below.)
2. Edwin Ardener (1975) in his article 'Belief and the Problem of Women' distinguished between dominant and muted models of reality by way of explaining the peculiar 'silence' of women in anthropological accounts of the field.
3. Fears of ghettoization and marginalization are perhaps quite legitimate when we see how disciplines tend to handle feminist concerns. The issue of whether this would give rise to the formation of a subdiscipline (Strathern, 1981) and the desirability or otherwise of such a development is another important question.
4. As Visweswaran (1996) observes, this division tends to replicate an implicit separation of theory from experience, also losing sight of the fact that ethnographies can be centred on women's lives without sacrificing a larger contextual reality and vice versa.
5. See John (1996: Chapter Two) for an interesting attempt to map the theoretical spaces between universalist and relativist extremes.
6. This point has had varied echoes, such as in the insider–outsider dichotomy which often translates into a contrast between a 'native' and a 'foreign' anthropologist and the 'authenticity' or otherwise of their accounts. Practices of fieldwork, as well as writing, however, could problematize such dichotomies. (See Narayan, 1993.) In India, the reflections on studying one's own society have been varied (see Srinivas et al., 1979; Beteille and Madan, 1975; Thapan, 1998). Some have dwelt on the moral/ethical problems in research (Omvedt, 1980; Karlekar, 1995), others on how

caste, education and other markers of the anthropologist influence the study (Srinivas, 1966). Yet others focus on the identity dilemmas of fieldwork (Narayan, 1993; Kumar, 1992) and the self–other questioning it entails (Ram, 1992). In all this, seldom has the *form* of the ethnographic account itself been subject to extensive scrutiny.

7. The writings of Metcalfe, Munro, Maine and Marx. For a review of these writings, see Dumont (1966), Srinivas (1987 and 1996b) and Thorner (1966).

8. Broadly, this polarization is into an empiricist approach to the village and a strongly culturalist one. As has been argued here, both fall short of engaging with the village on its own terms. It may also be observed that the more recent usage of terms like peasantry/peasant economy is beset by the same difficulty, that is, whether such terms can convey the distinctiveness of individual rural settlements.

Two

Of Gender and Space: An Outline

Most of our lives in society are anchored by a certain perception of (physical) space. Such a perception views space as an emptiness, almost as an open receptacle within which material objects are located. Rooted in our everyday experiences, this view assumes a certain hegemonic quality, often due to force of habit and perhaps the reinforcement provided by language. It has also formed the basis, frequently, for discourses and theorizations on/of space. Here, space is conceptualized as a void or an empty place, although paradoxically, it is this that lends substance to material objects. Put differently, space is thought of as an intangible, invisible backdrop, in spite of the fact that it is what enables the very definition of material bodies. I reiterate the latter, for, as is commonly known, one of the most important ways of defining objects is in terms of the spaces they occupy. In spite of this, space is pictured as nothing more than an unfilled gap or intervening distance between objects. So common-sensical and deep-rooted is this view that it becomes difficult to pursue any line of thinking that initiates a different understanding of space. And yet, it is precisely this task that I wish to embody here, even as I attempt to speak specifically of gendered bodies in and through space. By running a concept like gender alongside a category like space, it seems to me that one could open up crucial alternative conceptual axes within both realms. This opening up could, in turn, have implications for the ideas and discourses they serve to anchor.

To anticipate the discussion a bit, we could suggest that contrary to absolutist or essentialist views which have tended to fix and eternalize traits, space and gender are better approached as *sets of relationships* between phenomena, groups or persons, negotiated within certain given frames of reference. These relations are neither pre-determined nor programmed, but contingent, changing according to the context and the entities involved. In a manner of speaking, one could state that it is such a perspective on gender and space that will be outlined along the course of the chapters that follow. My more immediate objective in this chapter is to review some trends in socio-anthropological discourse, to see how it deals with, or deflects from, the spatial question. Indeed, as we shall see, even when there is an articulation (albeit indirect) of the dimension of space, seldom does it focus on its relationality with gender. Similarly, a look at the thematic concerns of gender studies in India finds it hamstrung between material and cultural axes in a manner that disallows any serious engagement with either the spacing of bodies or even, more strictly, the question of space.

Seeking a Spatial Perspective

My attempt here is to offer a rough sketch of socio-anthropological discourses, the intention being to size up and scrutinize the resources available therein for theorizing space and gender. The questions constitutive of sociology in the nineteenth century—essentially those of progress and development—were undoubtedly historical themes. The very formation of the discipline and the processes it sought to map revolved around the evolution and change of societies.[1] As such, it manifested an overt temporal orientation that effectively sidelined the spatial aspects of social life. It will not do, however, to argue thereby that the spatial dimension has been missing or entirely absent in sociology. A close reading tells us that the discipline does work with certain assumptions about space, though, by and large, these are not explicit and will have to be inferred through its other overriding concerns. For instance, representations of space, particularly in sociology, are often expressed in terms of context, where the latter serves as a mere locale or passive prop that allows history and social processes to play themselves out. Within anthropological studies, however, locale does figure as a central factor in defining the field, though it is never conceptualized in spatial terms. Instead, it is obliquely represented in terms of cultural differences, which are, in turn, seen as stemming from a certain structural arrangement, relationship or position (as in complex versus simple societies, or city versus country, or even core versus periphery formulations). In other words, a kind of superorganic conception of culture is made to stand in for the specificities of space.

To the above representations (or non-representations) of the spatial dimension we may add the tendency to collapse the spatial into the national. This occurs when social scientists including anthropologists proceed to posit the boundaries of a society as the boundaries of a nation (Agnew, 1993). This view of space as 'national' (as in references to an Indian or American way of life) has the obvious effect of displacing the specificity of the more 'local' or 'regional' variants of cultures. To be sure, none of these ways of speaking about space seem quite adequate, if our intention is to see how the social and the spatial are implicated in one another.

Given these condensed scenarios, it seems all the more important to attend to the tacit assumptions undergirding our discourses about space. Massey (1992) has in an insightful essay argued that across a range of disciplines space has been associated repeatedly with stasis, the status-quo, or a lack of politics. This view of space as an unchanging given, incidentally, is not a perception unique to social theory but also characterizes the discipline that explicitly addresses space and its object, namely, geography. Here, especially in its positivist incarnation, space has been conceptualized as an autonomous sphere with independent rhythms. Developments within radical geography, of course, have sought to challenge this view by arguing that space is largely a social construct, being constituted through social relations and practices; and indeed, by the 1980s, this formulation was further revised to take into account the other side of the picture, that is, the spatial construction of the social.[2] As Massey (1992: 70) characterizes this development, it involved the recognition that '...society is necessarily constructed spatially, and that fact—the spatial organization of society—makes a difference to how it works'. Subsequent writings of radical geographers have built on these ideas, enabling a re-conceptualization of space along more active lines.

It is only in recent years, however, that social theorists have begun to explore this issue seriously, in the process consciously seeking 'the reassertion of a critical spatial perspective in contemporary social theory and analysis' (Soja, 1989: 1). Often this has involved not a replacement of the temporal with a spatial perspective, but an effort to focus on the relations between the two. Such an elaboration of a spatial perspective has drawn (sometimes critically) on several, often diverse, ideas and strands of thought—the writings of radical geographers, feminists, Foucault's attempt to link space to the exercise and analysis of power, Lefebvre's Marxist analysis in urban studies, Giddens's time–space formula in his structuration theory, and of course structural anthropology's projection of a synchronic structure versus a diachronic understanding. Heterogeneous as these strands may be, each has, in its own way, facilitated the articulation of a spatial problematic, suggesting that space is not just physical form that may then be linked up with social structure but is very much the stuff of which social life is made. Most important of all, they have succeeded in focusing attention on the limitations of dominant representations of space, that is, of space as something abstract, immaterial, non-existent, a void that merely holds real objects. Pushing against an empirical notion of space as the intervening distance between two objects or points, it has today become possible to envisage a more theoretical axis that could foreground the import of spatial ideas on everyday life. It has become possible to deny that society and space are merely related externally; to argue instead for a focus on the embeddedness of life in socio-spatial contexts and locales. More pointedly, this kind of contextualizing and situating of social relations has created the conditions for a displacement of the universalized claims of social science. It has, amongst other things, enabled a perspective on 'gender (and other social) relations (as) constructed and negotiated spatially, and (as) embedded in the spatial organisation of places' (Duncan, 1996: 4). Indeed, within the discipline of geography itself, this move has also sought to re-politicize geographical concepts like space, place, the local, and so on.

The suggestive nature of these claims notwithstanding, there seem to have been only few attempts at systematically relating gender and spatiality. One early effort that does come to mind is the work edited by Ardener, *Women and Space* (1981). Asking how people map and classify their social worlds, the essays comprising this volume implore that spatial terms are indispensable in order to understand such a patterning. Societies generate their own culturally-determined rules for drawing boundaries in the real world, divisions that structure given modes of perception and social interaction. Terming these 'ground rules' and 'social maps', Ardener suggests that there are a multiplicity of such markers in our lives—whether it is in the delineation of political arenas, ritual spaces, or even public and private spaces within societies. An attempt is then made to bring out the interaction between space, on the one hand, and social organization, on the other, a task that is executed rather unevenly in the essays comprising the volume. While some show how people/ social identities are determined by the physical or spatial environment, others tackle how space itself comes to be defined by people through a series of conferrals of meaning and significance. More specifically, the book asks how women are situated in space, and to what extent this can be seen as a reflection of a cultural world-view and its ideas about sexual differentiation.

It is here that we may distinguish the concerns of the present work from the preceding collection. Rather than posing the question in terms of women's location in space (and thereby reinforcing an implicit separation between space and social worlds), I seek to invert the ground somewhat, to ask how space or spatial considerations figure in the constitution of femininity (and consequent sexualization)—both as a set of norms, ideas and ideals that are reiterated and as a practising ground that anchors and directs female lives in the community. Showing the way are Bourdieu's (1977) writings on the Kabyle, which provide a masterly demonstration of how space enters into the very constitution of social life and identities. His concept of the habitus reveals how a spatial classification is fundamental to one's social and cognitive map, providing individuals with a set of parameters within which to deal with everyday lives and situations. In Bourdieu's conceptual scheme, the habitus not only plays a major role in shaping and orienting our practical acts, but is also renewed along this course.

It shall be my endeavour to suggest that gendered bodies are at once spaced and spacing, being themselves constituted in space. To be sure, this recognition issues from the manner in which certain schools of geographers have articulated the space–society nexus and especially connects with their point that societies are constituted in space. However, while adopting the geographers' standpoint, my concern is also to transform it. This point would require some elaboration. Many geographers are, as Rose (1996) points out, critical of the wide use to which spatial metaphors are put in contemporary theorizing about subjectivity, location, power and the production of knowledge. It is maintained that there is a difference between 'metaphorical' space and 'material' space: that while the former, with its connotations of stable, inert, absolute space, provides a fertile ground for 'metaphoric appropriations' (Smith and Katz, 1993), it may be more important to focus on 'real' material space, including within this both geographical and social reality. Necessary as it may be to critically examine the attribution of qualities to space by theorists, I want to argue here against the theoretical dichotomization proposed by geographers. The point is not simply that the metaphorical and material can never be separated, but that the emphasis can frequently vary, necessitating a constant shift from one register to the other. This is particularly the case if we consider how certain cultural ideas of space feed off and into the bodily practices of women. Conceptualizing and embodying this socio-spatial matrix, as I intend to do, will require us to straddle both axes—metaphorical (or cultural) and material. Soja's emphasis on the need for elaborating a material frame of reference for space underlines, perhaps, the central thrust of such a task:

> *The generative sources for a materialist interpretation of spatiality is the recognition that spatiality is socially produced and, like society itself, exists in both substantial forms (concrete spatialities) and as a set of relations between individuals and groups, an 'embodiment' and medium of social life itself* (Soja, 1989: 120).

A spatial perspective must attempt to make clear how spatiality itself participates in the production of gendered bodies while also straddling its flip-side, namely, how such embodied persons negotiate their very social spaces.

A Materialist–Culturalist Impasse?

Such a move, of mapping the relationality and mutuality between gendered bodies and social spaces, has itself to be viewed, and assessed, in context. A focus on gender studies in India and its thematic concerns offers one mode of locating our questions and answers. Indeed, while the immense proliferation of empirical studies on women in recent years may have changed the contexts within which women are studied, it is apparent that the analytical frameworks adopted have not been substantially reviewed. The necessity for such a review lies not just in taking stock of existing analytical resources, but also in order to effect discursive displacements which could assist in re-mapping the domain of gender studies in India.

As even a preliminary reading will begin to reveal, gender analyses in India seem to obtain within a largely dichotomous conceptual grid, emphasizing questions of asymmetry, on the one side, and stressing qualities of womanhood, on the other. This grid frames a range of writings, whether from the nationalist period or from the post-Independence phase, even as it characterizes sociological studies on women. Chatterjee's (1989) analysis of the nationalist discourse provides a number of pointers in this direction. He argues that the charting of the social and moral boundaries of/for women in the nationalist discourse—effected through a delineation of the 'feminine' as an embodiment of the 'spiritual values' or 'essence' of India—has to be understood as part of the nationalist strategy of positing a dichotomy between material culture and spiritual culture. This strategy served to describe the material realm as the locale of colonial domination and struggle, while manouevering the spiritual realm outside the reach of colonial intervention. But, as Chatterjee notes, the burden of spiritual trusteeship on the inner, female realm nevertheless served to de-politicize and 'resolve' the women's question. The socio-political demarcations effected by the nationalist discourse—into home and world, spiritual and material, female and male—further facilitated a moral discourse hinging particularly on qualities of womanhood and feminine virtues.

With Independence, however, the women's question came to be re-surrected and posed anew in the light of the material inequities marking women's lives (Mazumdar, 1985; Mazumdar and Sharma, 1979). It is this latter context of a 'developmental' modernity that has defined a majority of the studies of women in India. A focus on the consequences of the new structural and institutional patterns generated by the develop-mental process, especially the gender inequities these give birth to, have been at the centre of such studies, with women's economic roles and employment receiving maximum attention. Purely descriptive accounts of female role conflicts, access to education, the problem of dowry, eco-nomic deprivation and appraisals of legislations on women have been among several other issues figuring within social science research (Niranjana, 1992, has the necessary bibliographical details). All these studies work with the premise, in itself indisputable, that an understanding of the material basis of women's subordination is vital in order to analyze women's status.

Running parallel to the above emphasis on gender asymmetry is another strand of thought more preoccupied with elucidating the meaning of 'femaleness' within the cultural–symbolic realm. This meets with aspects foregrounded in the nationalist discourse, but emphasizes the making of femininity in specific cultural milieus in the course of attending to the complexities of meaning and symbolism. The primary realms prob-ed by way of revealing the values underlying female roles in society are those of marriage, kinship and religion (Fruzzetti and Ostor, 1976). Approached as domains of culturally-defined social relations, marriage and kinship have here been linked to cultural categories and indigenous ideas about the person. Likewise, the focus on ideas of 'seed' and 'earth', which are approached as contrasting constructions of masculinity and femininity and as revealing the respective roles of men and women in society. Cultural ideas of femininity have also been thematized from the point of view of the social roles, values and behavioural expectations attached to the different stages of life (Khare, 1983). Equally, socialization patterns have been examined as transmitting some of these ideas and norms, which are, in turn, internalized by women in the course of shaping a self-image. In a similar vein, ritual symbolism has been seen as a powerful source of cultural ideas related to female nature and behaviour, especially conceptualizations of woman/goddess as both benevolent and destructive (Kondor, 1986; Tapper, 1979).

These conceptualizations of the feminine have tended to proceed independently of a focus on gender asymmetry, with hardly any effort being made to draw out their mutual implications. Of course, there are some early exceptions, as in Dube's (1986) important work relating the seed–earth metaphor to the asymmetrical rules governing the control and use of productive resources within patrilineal kinship systems. I shall presently consider two more recent attempts at overcoming the analytical divide. But considered overall, what is most striking is the dichotomy framing studies on women in India: between those emphasizing, on a certain materialist register, relations of asymmetry and deprivation and those, along a more culturalist framework, stressing qualities of womanhood and the feminine. The limitations of such a dichotomization are several. For instance, an attempt to represent womanhood in terms of cultural symbolism often lapses into essentialist analysis. This need not always entail celebrating innate traits of femininity, but frequently takes the form of assuming these traits as eternal, widespread and unchanging, all of which can culminate in an ahistorical view of feminine nature. Alternatively, the stress on gender asymmetry does succeed in incorporating a historical focus, but is unable to show how elements from a sociocultural realm infiltrate into the ordering of an unequal material world. From both these angles, then, a compelling account of the gender domain becomes impossible.

Such shortcomings have also to do with the fact that most conceptualizations of the feminine are hardly explicit about its bodily grounding. To some extent, this may have to do with an insistence within feminist theory itself—that gender definitions and identities be understood as socially constructed and independent of their bodily locus. This premise was based on an early separation made between sex and gender, a divide that has today increasingly been demonstrated as improbable, if not distortive.[3] Within sociological discourse, this relative silence about the bodily basis of gender was, perhaps, due to the dominance of the concept of socialization, which, in its own way, spoke of the learning of gender roles and expectations. Even amongst those studies that highlight the symbolization of the feminine, there are some which attempt to trace the socialization of the female through the stages of her life-cycle. And yet, the allusions here to the female body are nowhere near being spatial accounts, couched as they are in culturalist perceptions and expectations of the woman at diverse stages. They fail to ask, particularly, how women actually pattern these ideas in and through their bodies, as well as how bodies themselves become a focus for negotiating cultures, spaces and identities.

Rejoinder

The 1990s, if one were to go by the recent work produced, appear to be questioning both this bi-polar vision and the absence of the body in gender studies. I intend to consider two texts that are emblematic of this trend—Kalpana Ram's *Mukkuvar Women* (1992) and Karin Kapadia's *Siva and her Sisters* (1996). There are interesting similarities between the two—both are based in rural South India (more specifically, Tamil Nadu), both focus on women (those belonging to the lower castes, albeit different castes), and both seek to combine an examination of the politics of the household with that of the workplace or larger community. The parallels do not end here. Both also seek to engage with certain dominant conceptualizations of Indian society, especially that of caste hierarchy and the (Dumontian?) idea of cultural consensus, and try to offer a different reading. In describing itself as 'a study of social change and capitalist transformation' (p. xv), Ram's study is also challenging Marxist economistic views on Indian society by showing how the cultural construction of gender occupies a central place in capitalist change. Kapadia, by centralizing women's lives and experiences, focuses on the reverberations between gender, caste and class as three axes of identity. While each work offers a detailed ethnographic account of the respective communities chosen and conveys the uniqueness of particular circumstances, there are some general points on which they converge. Thus, the focus on puberty rituals, kinship and marriage transactions, religion and possession on the one hand, and the effort to relate gender and production politics on the other, is an attempt obviously at overcoming the analytical divide between materialist and culturalist emphases.

Working broadly with the notion that a labouring body is also a gendered body, Ram and Kapadia nevertheless present contrasting ways of bringing the material and cultural axes together. The former's ethnography self-consciously seeks to scuttle the polarities between these axes, and 'plac(es) culture and sexuality at the centre of capitalist transformation and the new division of labour between men and women' (Ram, 1992: 234). For her, an important way in which avowedly cultural discourses of sexuality and gender could be infused with a certain material reference is by taking recourse to the labour process. The body is a crucial indicator that 'mediates between economic disciplines of the labour processes and the sexual disciplines of gender identity' (Ibid.: xvi). On the other hand, Kapadia does not self-consciously centralize the body, but highlights instead the everyday forms of life and resistance among the women of the Pallar and Muthurajah castes. She draws on their concepts, rituals and daily discourses in order to maintain that studying the dynamics of class and caste processes would be impossible without a gendered analysis. Each in its own way serves to draw out the material resonances within the gender domain and are valuable for this very reason.

An even more recent attempt at straddling the culturalist and materialist divide in gender studies can be observed in the elaboration of indigenous ideas of the gendered body in South India. Busby (1997) posits that gender identity (being based on bodily difference) is fixed and stable, and that it is focused centrally on the reproductive capacities of the people involved. Drawing on her ethnography among the Mukkuvars in Kerala, she argues that the definitive qualities of maleness and femaleness emerge only through a comprehensive process involving their transactions with each other. Such transactions and exchanges get further cemented in the course of their economic, sexual and kinship relationships. As Busby notes, '(gender)...difference and separation are dependent upon co-operation and interaction, and women and men become most fully gendered in their cross-sex relations with each other' (Ibid.: 267). Busby's perspective on the conceptions of the gendered body make her relate these to ideas of personhood in different cultures, her interest being primarily in a comparative axis involving diverse non-western cultures.

As a response to the bi-polar vision marking gender studies in India and the absence of the body from that discourse, the points that I wish to explore, and develop gradually, are of a slightly different order. It could be inferred from the above attempts to initiate a dialogue across materialist and culturalist perspectives that they seek to re-materialize the gender domain, and do so by working through questions about labour and resources.[4] Without denying the legitimacy of such starting points, I venture here a different axis of problematization. This is based on the realization that the materialist and culturalist characterizations of the gender domain are not two unrelated problems that need to be tackled separately, but, in a sense, underwrite one another. My suggestion is that a transcendence of the culturalist–materialist polarity can only be achieved when the body itself comes to be inserted into our moments of investigation. Focusing on space, and the body's relation to space (social and physical), could provide us with a tool for effecting this inscription. Building on the feminist argument (also adopted by Ram) that the body is the central material anchor for discourses on gender and sexuality, we should go on to ask how bodies—particularly female bodies—inhabit and negotiate space, as well as how such spaces are defined, and how space figures in the daily lives of women. To pursue this line of analysis would be to re-materialize the gender domain of course, but would also inscribe a materiality previously neglected or even taken for granted: one that has to do with the spatiality underscoring women's everyday lives.

In the course of charting such a terrain, however, I must reiterate that this book actively strives to depart from the dominant trend of utilizing a purely monographic format. Confounding demarcations between the field, its representation and analysis, the interpretation in this work enacts, in some ways, the dilemma of writing ethnographies in and of the everyday. To be sure, considerations of narrative remain on the surface, but the chapters are also attempting to condense thoughts about gender and space in the Indian context.

Notes

1. See generally Krishan Kumar (1978). The introductory chapters of Giddens (1982) are also illustrative.
2. This was a point also argued by Lefebvre (1991) much earlier in his *The Production of Space*.
3. I shall refrain here from further comment, and will engage this ground later in Chapter Eight. Between our perspectives animating gender and the postulation of a range of phenomena held to issue off (or inflect into) this realm, there is a ground to be traversed. In its largest aspect, our ethnography functions as this middle ground— an order of space measuring an expanse of names, points and limits.
4. Busby (1997) perhaps is an exception, but even here the claim that transactions are central to the substantialization of gender detracts from the significance of bodies as gendered.

Three

Spatial Registers

Fieldwork carried out in a village cluster in Karnataka some years ago anchors and gives body to some of the observations of the preceding chapters. The cluster, which I have called Aaladapalya, envelops a number of settlements—two large villages (Hulaguppe and Doddaaluru) and four associated hamlets (Shivapura, Arasapura, Budkepalya and Chikkuru) spread over a radius of 2 to 3 kilometres.[1] At the hub of these villages is a big Banyan tree (*aaladamara*) that serves as an identity-marker and a mode of characterizing the entire locality—not just in terms of geography, but even culturally. The goddess Kunigalamma is the presiding deity of Aaladapalya as well as other villages (thirty-three in all) in the vicinity.[2] With the exception of one hamlet (which is a relatively new settlement), the main villages are fairly old and can trace their genealogies back to nearly four or five generations.

Despite their seeming proximity to Bangalore city, these villages are not economically prosperous, exhibiting, in a way, the underside of economic development in the region. The key unit of production is the family/household and agricultural production is either for subsistence alone (in the case of marginal farmers) or for commercial purposes. The village cluster is also involved in a webbed network of ties with surrounding villages, towns and the city. Though there is a distinct separation between such a region and the urban centre in terms of economic organization, the influence of the latter is felt at various levels. For instance, proximity to the city has opened up work opportunities that were non-existent a few decades ago, altering relations between caste groups. Wage labour has wrought changes in caste relations and in the agricultural realm, gradually replacing 'personal bondage' with 'de-personalized labour power'. Indeed, in spite of the village cluster having its own socio-cultural specificities, the fact that it is drawn inexorably into wider processes of economic change linking it with a pervasive commercial ethic has made it impossible for the community to remain unaffected. Rather than examine the economic or material implications of this, my endeavour will be to see what underscores the definition of rural spaces, as well as to examine how such apparently indeterminate contours of rural space inform the bodily practices and orientations of women. In many ways, this chapter marks the contours of what is to follow.

An Initial Evocation

A widespread tendency in delineating a village is to adopt a geographical description that usually overlaps with administrative demarcations of territory. Such an approach tends to take village space as a territorial given, often unwittingly obfuscating the modes in which space is socially ordered and lived. The attendant dissociation between the social and spatial registers in official/administrative definitions of village territory could lead to rather diverse delineations of the village. Thus, if we pursue the official definition, all the villages and hamlets in the cluster would belong to the same administrative unit, coming under the Tavarekere Hobli Panchayat, Magadi Taluk, Bangalore District. At the time of the study, the earlier village panchayats had been officially dissolved with the intention of bringing the region under the mandal panchayat programme. Besides, deeds of landownership and other records pertaining to Doddaaluru (the large village) and Arasapura (a hamlet) were maintained at the village accountant's office at Doddaaluru. The other three hamlets, along with the other large village Hulaguppe, formed the Hulaguppe Panchayat. They were described as the *dakhale grama* of Hulaguppe, that is, those villages whose land and revenue records (*dakhale*) were maintained at the larger revenue unit.

Such an official/administrative delineation of territory is hardly indicative of the modes in which people experience and negotiate their village spaces. For instance, people overwhelmingly conceive of village boundaries (*uru yelle*) in terms of the shrines of their guardian deities (as in Doddaaluru, where Veerabhadra, Nandi and Mulkatamma mark and guard the frontiers). In some cases, boundaries are determined by natural factors like hillocks and streams (for example, the separation of Budkepalya from Hulaguppe or Chikkuru). Also fairly common is the inclusive definition of a whole region, as when referring to the thirty-three outlying villages as all belonging to the goddess Kunigalamma's domain. Emerging along a different register is the significance of affinal versus natal ties in defining one's village, with women often meaning their natal home when they spoke of *nammuru* (my/our uru). Caste too played a role in definitions of village space, demarcating the village into *urina olage* (inside the village) and *urina horage* (outside the village). Exploring the significance of such an inside–outside matrix is a task that I will recurrently address throughout the book. What needs emphasizing here is that in addition to geographical and territorial considerations, people's definition of village boundaries and their negotiation of space bring into play a number of elements such as ritual, caste, kinship and, of course, gender.

Thus framed, the connotations of the administrative category of *grama* seems much more restrictive than the culturally resonant uru. Daniel (1987), in fact, has pursued this point. His focus is on the 'cultural world' of Tamil villagers, where he attempts to develop an understanding of the substances (or properties) comprising bodies, home/hearth, village, and so on. He stresses, above all, that there is a constant mixing and transformation of such substances, as for instance, when territorial substances mingle with bodily substances in the constitution of personhood. Such a semiological analysis of the cultural world of villagers, although interesting, is not pursued here. Nor am I, for that matter, trying to draw exclusive attention to the cultural specificity of the village by raising the question of place. Rather, I am interested in exploring the point of imbrication of the social and the spatial, as bodied forth in everyday practices and perceptions, as well as in showing how spaces themselves are gendered. The idea is to map the process of the delineation of space, as negotiated through the female body, where descriptions of community and identity are mediated through rules of movement and female morality. To reinscribe the spatial into the social is thus to embark on a fresh narrative of the village as social space, with the woman being the key representing medium.

Drawing upon suggestions thrown up within the village cluster studied, we could, at one level, indicate the dimensions of spatiality that inform the bodily practices of women within the community.[3] At another level, we could also draw attention to how the body, and the modes in which it inhabits space, itself comes to be deployed as a medium through which the 'female' is constituted, even sexualized. The mediations suggest that the socio-spatial parameters of women's lives hinge on a delineation of what is known in the vernacular as olage–horage. Whether one is talking about matters as diverse as work, family, caste group, quarrels, 'proper' behaviour, a range of women's activities, or quite plainly interaction within the village, all these are described with recourse to the vocabulary of olage–horage.[4] These spatial ideas, strongly embedded in the perceptual schemes of people, emerge as the principles orienting their daily practices, the axes along which the world is ordered into one's own or other, female and male, familiar and strange, proper and improper, and so on. Yet, these lines are neither rigidly nor essentially defined, since the parameters of what is inside (olage) and what is outside (horage) itself keep shifting. The following sections attempt to unravel these notions, offering an outline of the registers of spatiality and femininity/sexualization.

Siting the Body

The ensuing account of the domains and activities of women, as well as a consideration of the ways women speak about their lives, offer some indications of the modes in which bodies and spaces are gender-marked and sexualized.

The central preoccupations of women in the village cluster—across differences of caste and economic standing—are the needs, even the survival, of their families. The household, in more ways than one, is located at the centre of these women's lives, being both the object of and the locale for a large chunk of their daily activities. To be sure, in the course of their activities, women also negotiate a variety of situations and circumstances that may even take them beyond the physical space of the household. Starting with a woman's typical household activities, one can see how her movements are choreographed by certain implicit cultural rules governing the use of space. For instance, most of them in the course of their household chores rarely venture beyond certain conventional boundary markers separating uru (village) from kaadu (though literally forest, it refers here to the uncultivated wild spaces surrounding the village).[5] Even in carrying out their tasks of collecting fuel and grazing sheep or cattle, women go in small groups and follow fixed and familiar tracks immediately around the village, always keeping the settlement in view or within shouting distance of one another. There is also a strong reluctance to enter groves and bamboo thickets, or cross surrounding hillocks, fearing an exposure to undesirable people, spirits or places. Such perceived dangers define what is out of bounds for women, and in doing so, reinforce the security and informality of the village, as an inner space that is one's own.

Having a telling impact on women's use of space are the variations that are introduced into the work-day patterns of women. Among economically worse-off groups—who are also largely lower castes and usually landless—a woman, in addition to domestic chores, also undertakes wage labour. Her options, however, are limited in comparison to men and she looks for employment as either an agricultural or a construction worker within her village or, at the most, at the brick kilns situated in the spaces bordering her village and the next. This is not only because of her domestic responsibilities but also because, as we will see, transgressing these spaces, even in search of subsistence work, is often seen as casting a shadow on her and her family's moral standing.

Among better-off families belonging to the dominant castes, particularly the small and middle farmers in the village, women withdraw from wage labour, though they continue to work on family fields in addition to doing the household chores. Yet, the widespread assertion by several women like Kamalamma, a Vokkaliga woman, that they do not 'work outside' suggests that they perceive this space, that is, the family land and their work on that land, not as something external but as an extension of their household chores. Sowbhagya's perceptions of the nature of women's work establishes this continuity:

I'm not denying that a man's work is important...but a woman's work is the main thing. We not only have to toil in the fields, but also in the house...attend to the children, attend to the house, do the cooking, and when your man goes to the field, go there and help him too.

In spite of recognizing this double burden, the space of the household is nevertheless expressed as the sole and primary responsibility of women. Sharadamma, a Lingayat woman, underscores this sentiment:

Who will cook and care for the children and home if not the woman? What will happen to the household if we don't? The dharma *of men is different...they only do outside work.*

This identification of women's interests with the inside—olage in the local idiom; here the household or hearth—is most strongly expressed among women of upper castes who do not seek wage labour and for whom the household becomes the sole preoccupation. Their access to economic resources is limited and their seclusion induces them to concentrate mainly on maintenance and reproductive tasks. To that extent, their mobility is restricted and their visibility also decreases, since their interaction is largely confined to the caste neighbourhood. As Rajamma pointedly observed:

> *Our menfolk labour outside and provide for the household; as for us, we look after work inside the house. We don't need to go out for any-thing.... Besides what business have we going near places where men are?*

Such assertions underline the significance of space as structuring a woman's world, an overarching principle which is so taken for granted that it often remains invisible in analysis.

The foregoing account marks an initial two-tiered understanding of women's location in space (to which we shall soon add other dimensions). One is the perception of the village as 'safe' for women; the other, by further localizing this space, has to do with the household or hearth as an inviolate and intimate realm for women. With these preliminary indications of how such spaces as uru, household and field are gendered and circumscribed, let me intersperse a series of random expressions of people in the village and draw attention to another crucial discourse folded into this process, namely, a discourse on female morality and sexualization. This would help account for how rules regarding the use of space are themselves constitutive of perceptions of the female body.

A reference was made earlier to the variations introduced into analysis by attending to the work dimension. Work opportunities for women outside the home are structured in definite non-economic ways, with the nature and place of work being major factors that are taken into consideration. The workplace of female labourers is invariably near and familiar, since travelling long distances to work unaccompanied not only carries the fear of what is alien but is also seen as 'wrong', bearing the stigma of moral suspicion. Women who transgress their habitual, 'assigned' socio-physical spaces run the risk of being labelled of 'loose virtue', and are subject to strong censure by older members. In some cases, women were forced to discontinue going to work outside their village, since elders considered it 'unsafe' and 'immoral', a blot on both family prestige and women's honour. The opinion expressed by an elderly male, in response to women who worked all day at a seasonal fruit-pulping unit several miles away, seems typical of such a sentiment:

> *The point is not one of simply wanting to earn money. True, nothing will happen to those among them who are 'straight', and can work and earn with their wits about them. But there are others who are*

spoiled. It's not right, going out like that.... Once you do, who can say whether you are upright or not?

This concern with regulating women's sexuality/morality by confining them to 'safe', legitimate places, such as inside the house or inside the village, does not go unquestioned. As Sharada, a young Vokkaliga woman, opined, partial transgressions of women's space become inevitable in the face of dire economic necessity:

If we are totally confined to the house, then what will happen when there is some trouble? Now that Mahadevamma has been deserted by her husband, she is back at her mother's place with the little children. But how long can they support everybody? And can she just sit at home? It is she who has to feed her children now.

The delineation of inner spaces, such as the household and one's own uru, as safe, legitimate places for women strongly suggests that these spaces are also being projected as intensely moral realms. Everyday expressions, such as that of a mother chiding her young daughter, 'Why are you standing at the door, you slut? Get inside the house. Don't you have any work to do?' or 'Can't you come back home immediately? What work do you girls have on the street?', carry implicit references to shame, honour, chastity, female sexuality and its regulation. Yet, there are residual voices that co-exist with these ideas of the inside as an inviolable space for women, which puts the onus on one's good conduct, rather than one's surroundings, as decisive. According to Rudramma:

Girls going out is not wrong in itself. It all depends on how we conduct ourselves. If we are straight, who can do anything to us? Gold will shine and be pure wherever you throw it. But some others are not like that. However much one tries to hide them from the public eye, they will spoil. They needn't even go out...for those who are bound to spoil will spoil even if sitting before the stove.... That inside is safe and outside is not is all false. Our honour and shame remain in our own hands.

Although there is an apparent inversion of the 'inside as safe–outside as unsafe' perception here, a closer scrutiny would reveal that not much has changed; what is being posited is only one more layer of the inside—an interior moral sense, if one were to have it—rather than the association of morality with physical spaces alone. The crux of the matter remains female morality and its regulation.

Further Partitioning of Socio-spatial Matrices

To these paths and points along which women's spaces and identities are marked can be added another dimension—that of caste values—as the major axis along which issues of space and morality are addressed. There have been several competent analyses of the centrality of the proper channelization of women's sexuality to the maintenance of caste purity (see Das, 1988a; Ganesh, 1989 and 1993). I am interested, however, in pursuing a slightly different line, highlighting aspects such as women's assigned role as major custodians of caste values (see Dube, 1996), and also indicating how the imperative to maintain caste boundaries complicates the ways in which women (and men) inhabit space. Typically, observance of rules of interaction between castes are indicated through such statements as 'We don't eat at their place', premised on ideas regarding the superiority or inferiority of castes. A majority of the upper caste women, in particular, are punctilious about caste rules in their daily lives, so much so that it is remarked: 'What do we have to do with them (the lower castes)? They lead their lives, we lead ours (*avaradu avarigaayitu, nammashtakke naavirutteve*).' However, a closer scrutiny discloses several points of everyday contact between castes. One aspect that links together higher and lower castes is the sphere of work. The former largely depend on the latter for most of their day-to-day labour requirements, be it work on the fields or other chores. There is nevertheless a clear-cut spatial distance maintained between them: the lower-caste person will never step over the threshold (*hostilu*) and enter the upper-caste house. Even a woman assigned the task of cleaning grain, for instance, sits on the verandah (*jagali*), a liminal space that is neither as public as the street outside nor as secluded as the inside. Within the space of the village, everyday practices follow a certain routine which, in effect, ensure that caste rules are not sidelined.

Some women (largely upper caste) pointed out that despite this, caste rules are often flouted by men. According to them, men are not only more mobile, moving in and out of the village, but also deal with a range of different people and situations during the course of their work. This creates situations wherein pollution rules cannot be adhered to meticulously in the 'outside' realm. Thus, caste rules get suspended until an individual's re-entry into his uru. This 'slackness' on the part of men (in the vernacular rendered as *bigi illa*[6]) is seen as placing an additional responsibility on women in observing caste rules and conforming to caste behaviour. As Shivalingamma observed:

*Actually, it is women who follow caste restrictions so much.... If she
doesn't, then she is condemned outright, and people will worry about
caste purity. But whenever a man neglects caste rules, he is merely
reprimanded, or ridiculed as ignorant and brash.*

However, a Holeya (Dalit) woman added another dimension to the
consideration by asking:

*What caste are we talking about? 'We don't eat here, we don't marry
there', they say. Yet, a Lingayat can keep a Madiga woman, or a
Brahmin a Kuruba woman. Nobody says anything to them.*

What this suggests is a clear unevenness in the application of caste rules
and an underlying power dimension in the traversals of the space of
caste.

Without doubt, the observance of commensal proscriptions is one arena
through which one can trace the socio-spatial ordering of caste interaction.
Guarding against the violation of these rules is a dominant concern among
the Brahmins and Lingayats who see themselves as superior (*naavu
uttamaru*) and do not eat at Kuruba households or at any other household
below them in the caste hierarchy. The Kurubas do not eat in a Madiga
household, whereas the Madigas and Holeyas (divisions among the Dalits
into left-hand and right-hand castes respectively) do not accept food
from each other. Just as food represents one axis of socio-spatial divisions
between castes ('we don't eat at their house'), another significant way
of charting spaces in and through caste matrices is to say that there is no
'give and take' (of brides) between them (*kododu thogollodu illa*). While
for the upper castes, both of these facets—that is, the acceptance of food
and brides—are crucial, the Kurubas, Dalits and Budabudikes often stress
the latter in delineating their relations with others. In emphasizing that
there is no 'give' and 'take' of brides between them, what is being specified, note, is the space of the sub-caste and not the caste as a whole,
alerting us to another level at which social spaces are being partitioned
into olage and horage. When a Kuruba woman, for instance, describes
herself as Haalumatasta and says that they don't marry the Jenu Kurubas
(another sub-caste), it is her own sub-caste which constitutes the inner
space, the other Kuruba sub-castes perhaps forming the first line among
a circle of successive outsides.

Among the Budabudikes and Dalits too, sub-castes are very much
part of everyday social reality. And yet, even as these lines of distinction
are reiterated and used as broad markers in negotiating caste interaction,

people across castes also aver that changes are slowly creeping in. Among the most significant changes has been the increase in cases of inter-marriage between sub-castes. What was unheard of a generation or two ago, is now socially accepted. As a Vokkaliga woman, citing the instance of a wealthy landowner who had just married his daughter off to a rich man from another sub-caste, put it:

> *Once the big people do this, what is left for us? ...in (fixing a marriage) in this day and time, one must not look at caste or lineage (*jati-kula*) but see whether the family is well-placed.*

Thus, while the contours of the sub-caste seem to be undergoing some change, its effect has been to re-trace and consolidate caste identities and boundaries themselves. While it would be premature to draw conclusions without more in-depth explorations, it could mean that caste is replacing sub-caste as a significant inner space.

Although caste interaction is underwritten by such notions of space as one's own (*nammavaradu*) and other (*aa kadeyadu*)—or even inner (olage) and outer (horage)—it should be noted that the parameters of these realms are not fixed. Nor are they uniformly the same for men and women at all times. The olage could be either the family, kingroup, sub-caste, wider caste group or uru, depending on the gender we are talking about (whether men or women) and its relation to the larger entity (for instance, town/city, or other villages) it is being compared with. With particular reference to caste spaces, it appears that women are more concerned about socio-spatial rules as they impinge on an inner domain (be it the family, neighbourhood or sub-caste). This is not to suggest that men are unconcerned with the 'inside' (in this case, the context of households), but one can maintain that the forms within which they articulate and reiterate caste distinctions are different, somehow bordering on outer rather than inner spaces. Seeking an exclusive hold over ritual spaces within the village could be one indication of how castes and 'its men' seek to express mutual relations of superiority and inferiority. Mulling over this possibility, I came across this entry in my field-notes:

> *The Basava temple in Hulaguppe has received a face-lift. Previously, it was an open one, with an inner door protecting the Nandi and an open jagali which was used by villagers—from school kids, to bus passengers, to card-players. Now a grill has come up all round, locking the veranda in as well. I remark upon it to the old Lingayat woman who minds the shop close by. She replies: 'The Madigas used to sit*

there—spit all round, play cards and dirty the place. If we said
anything, they would retort—'does this god belong to you alone?'
...That's why our men decided to enclose it...'. Like elsewhere, the
Madigas here are not allowed inside the temple. Only during festivals,
some of them offer worship, but by standing outside itself.

What I have attempted thus far demonstrates the mobility that attaches
to the demarcations of olage and horage, and the differential negotiations
of these spaces by women and men within and across castes. Such con-
densations of the fluidity of socio-spatial matrices, it seems to me, could
be invested with significant theoretical functions as well. Most importantly, it serves to counter the implicit tendency to view space as a mere
context or setting for interaction, as a passive dimension that is determinate and unchanging. Space and society are not only highlighted as
mutually constitutive, but the analysis goes a step further in complicating
this terrain by showing how femininity and sexualization come to be
necessarily implicated within the socio-spatial. The point is not so much
that these processes of circumscribing the female body are occurring in
space; rather, that femininity/sexualization are, at once, the product of
and the medium through which physical, moral and cultural spaces are
marked.

Indeed, what is referred to as the olage is set up as a fundamentally
moral realm, preoccupied with issues of women's honour and family
prestige, on the one hand, and caste idioms of purity and pollution on
the other. This space, which coincides by and large with the domains
inhabited by women, is never only the household, but includes the caste
group and/or the village at different contextual junctures. Woman, and
the manner in which she inhabits space, becomes a key to preserving the
parameters—physical and moral—of the group: one, in that considerations of protecting female honour and regulating sexuality not only
inform the rules underlying women's use of space but are also the hinge
along which groups—whether these be households, castes or the village—
draw their boundaries; and two, in that women themselves are often
central actors in this process of cultural reproduction. In delineating the
significance of spatial contours thus, we are far from freezing interaction
into a static mould. Indeed, it is only when one accords a recognition to
the flow of the temporal that the seams of the spatial become visible. In
other words, the olage–horage delineation, while normally a grid that is
taken for granted, can become a point of reflection and refraction during
periods of flux or moments of transgression by individuals. It is in such

contexts that normative codes come to be reiterated and socially reproduced. There is yet another sense in which one can underline the importance of space and gender. This is by asking whether, in demonstrating the 'location' (*sic*) of women in particular socio-spatial contexts, one can re-address the question of female agency. I engage with aspects of this question in Chapter Six.

While the contents of this chapter have only tried to gesture towards the complexities attached to the socio-spatial matrix, it would not be an exaggeration to assert that spatial considerations are central to the articulation of a logic of everyday life. While there are situations in which these find verbal expression, more common is their operation through a range of practices. Examining some of these practices will be the task of the subsequent two chapters.

Notes

1. Fieldwork was conducted in a village cluster outside Bangalore between 1988–89. A socio-material profile of the village cluster has been provided in my dissertation 'Symbolic Meaning and Rural Social Structure: A Sociological Study of the Construction of Femininity', Bangalore University, 1994, and also summarized in the appendix to this book. The village cluster lies about 30 kilometres west of Bangalore. The settlements are multi-caste, with households depending largely on agriculture, either as independent farmers or wage labourers. The major castes are Lingayat (a dominant upper caste), Dalit (who at times describe themselves as Adi Karnatakas, less frequently as Scheduled Castes, but more often in specific sub-caste terms such as Holeya and Madiga), Vokkaliga (a middle peasant caste, though not uniformly prosperous) and Kuruba (an impoverished, marginally landed, lower caste), with a smattering of service castes. Note, all the names of the villages have been changed.

2. The one exception to this is Doddaaluru, which also does not participate in the annual jatre of the goddess. The village, however, propitiates Mulkatamma, who is said to be one among Kunigalamma's sisters.

3. While the variables of caste (and class) are undoubtedly of determining significance in that they introduce certain complexities into women's use of space, it must be underlined here that across these variations is an underlying homogeneity in the modes in which the spatial idiom is persistently used to refer to issues of female morality, shame, honour and rules of movement. This raises the larger question of whether, and how, spatiality can be conceptualized as a sort of structure grounding even caste (and class) and entailing a quasi-determining influence on their negotiation by gender.

4. The axis of olage–horage may be roughly rendered as inside–outside, though the latter may not entirely capture the nuances and shifting boundaries of these spaces. Discern of course the considerations to follow in the rest of the book.

5. This point assumes special significance, given the assertion from some quarters of women's proximity to nature, a standpoint which also folds into the tendency to

persist with the nature/culture distinction when speaking of women (cf. Ortner, 1974). By attending to the actual movements of women in and around the village, the argument implicit here is that the body–space encounter is layered with references to 'nature' and 'culture' in ways that are not always strictly separable. See MacCormack and Strathern (1980) for an examination of the cultural specificity of the nature–culture divide and the complexities attendant on its mapping in diverse cultures.

6. This phrase—which could be rendered as loose, lax, or more literally as 'not tight' enough—is invoked in other contexts as well, for instance, in the course of suggesting a 'carelessness' in money matters. What is striking, however, is that while laxity on the part of women often becomes an issue of morality, in the case of men it tends to be condoned as an individual trait, an oversight, or as an understandable fallout of being male.

Four

Bodily Matrices

The delineation of olage–horage, it should be stressed, does not constitute an abstract principle of space, but rather is accomplished through socio-cultural practices. Embedded in the perceptual schemes of people, their deployment constitute the axes along which the world is ordered into one's own or other, female and male, moral and immoral, and so on. As the framework orienting the everyday practices of men and women in the region, olage–horage is also the idiom in which symbolic expressions of bodily experience are cast. The association of olage with women, for instance, brings with it assurances of safety, morality and honour, prompting the suggestion that it (olage) be read as a bodily metaphor encoding certain cultural conceptions of the feminine.[1]

The predominant focus of this chapter will be on women and the cultural beliefs and practices impinging on their bodies. It supplements and extends ideas contained in the foregoing chapter by highlighting how the socio-cultural realm spatializes specific norms of female conduct, which, in turn, both regulate and are consolidated by the actual bodily practices of women. The system of shared norms orienting speech, perceptions and behaviour—dispositions, in short—within the community would be central to this understanding.[2] Being shaped by and constitutive of more regular contexts of practice, this body of norms and dispositions represents a matrix of sexualization crucial for an understanding of certain 'practices of the self' of women. Focusing on everyday contexts, this chapter consolidates into a mapping, how the female body is perceived, presented and spaced.

Innu Hennadalu: *The Significance of Puberty*

What does it mean to live as a woman? Or, more specifically, to live within a female body? One way of answering this question would be to turn inward, provide personal chronicles of women's experiences and perceptions, and so on. Another way—which is what is attempted here—would consist in turning to the culture speaking about the feminine and the female body, as also its symbolic (read, life-cycle) manifestations. The body, even as a biological entity, is never simply a given, but is always mediated through the socio-cultural, such that even one's own experience of the body is invariably through this register. The moral injunctions and norms that are seemingly part of the social realm are in fact inseparable from a culture's imaging of the female body.

Everyday life in Aaladapalya is held together by a closely knit system of norms and values, a number of which impinge on the behaviour of women. Prominent among these are the concepts of 'shame' and 'honour'. The former pertains to a comparatively individual-centred feeling, while the latter brings into play various domestic or familial groups and their reputations as well. Women's honour is unquestionably related to the specification and maintenance of group boundaries, although this is a line of analysis we will not pursue at the moment. Instead, we will focus on how issues of morality and honour govern the experiences and lives of a girl (or woman) within the community, pre-scribing the spaces available to her as well as indicating how she is to function within them.

The onset of puberty marks a critical turning point, both in terms of societal perception and in terms of an altered self-orientation to the body. Though it is the ritual of the first menstruation that marks and formally announces the maturity of the girl, ideas of shame–honour (*mana mary-ade*) and womanly restraint have, by then, already been internalized by the girl. Essentially, the ritual itself commemorates a 'fertile virgin', one seen as equipped for childbearing and childcaring tasks. Certain crucial cultural images and categories, therefore, mediate in the conceptualization of the female body. The mediation is so complete that women come to relate not so much to the body, per se, but to the cultural ideas underlying each change (or stage of life) and embodied in their speech about bodily processes.

Even though women talk about it amongst themselves, there is no direct talk regarding menstruation (or even any explicit instructions) between, for instance, mothers and daughters. This is perhaps reinforced by the thought that the mother should not be the first to hear of her daughter beginning to menstruate. Children usually learn about it by overhearing instances of somebody else 'coming of age', or sometimes having observed the segregation of (older) menstruating women, either at home or at the village water-tap. Their understanding is tied not so much to their own experiences of bodily changes but to ideas of impurity, which are perhaps echoes of the more widespread cultural perception of the female body as impure, periodically recovering a semblance of purity after each menstrual cycle. The feeling that the mother should not be the first to hear about her daughter's menstruation is also linked to the other problems that could arise with the girl's emergent sexuality. Some informants rationalized this by saying that it serves as a dreaded reminder of the duties of parents regarding their daughter's marriage.

The readiness of the girl for marriage and childbearing is variously brought out in the expressions that convey at once the cultural status of the girl and the way her life is spaced. She is said to have 'grown up' or matured (*doddavaladalu*), become wise (*buddhi banditu*), or is described as 'sitting out' (*horagadalu*, referring to the practice of segregating the menstruating woman from daily activity spheres). More revealing is the usage *innu hennadalu*—'she has now become a female'—suggesting both the relatively asexual purity of the pre-pubertal girl and the equation of womanhood with childbearing. The latter aspect is more specifically addressed in phrases like 'now she is ready' (*innu tayaradalu*), 'her body has filled out' (*mai nereyitu, mai tumbikonditu*), or 'she has flowered' (*hoovadalu*, indicating parallels with a tree which bears fruit consequent to the flowering stage).

These ideas are implicit in the puberty ritual, a symbolic performance embodying certain social conceptions concerning 'girls, women, femaleness and womanhood' (Winslow, 1980: 605). Signifying a transition from one cultural status to another, the puberty ritual is a highly visible rite of passage—a spacing, if you will—which both affirms and facilitates the social passage of the girl. Throughout the village cluster, the first menstruation of every girl is a public ritual occasion, signalling to the community the actual transition made from girlhood to womanhood. Its importance also derives from its significance in influencing the future marriage of the girl. Indeed, in quite a few cases, marriage proposals among kin are initiated and/or finalized in such contexts. The separation of the girl from her previous social status is reinforced in physical terms by a parallel geographical shift, albeit temporary. This is effected by the actual or symbolic construction of a grass-hut or shelter (*soppina gudisalu*) just outside the house, under or near which the girl sits for three days.[3] This aspect of the ritual is also indicative of the 'liminal' status of the menstruating girl yet to be incorporated into womanhood. More strikingly, it reflects the dialectic between the natural (the leafy hut as signifying nature or the wild) and the cultural, where the cultural not only accords recognition to natural changes in the physical body but also seeks to transcend nature in specific ways.

Though the first menstruation is in itself an auspicious event, the menstruating girl herself is seen as 'polluting'. In fact, the entire village observes a 'pollution period' in matters relating to ritual during these days, with events like the annual village festival not being held during that time. Menstrual blood is perceived as 'dirty' (*holasu*) for some, as 'dangerous/harmful' (*kettadagutte*) for others, and a matter of shame for women. The spatial segregation imposed during the first and subsequent menstruations serves to curtail the normal social interaction of the girl. Nor can she participate fully in the customary activities of the domestic sphere. She is given a separate plate, mug and mat, and there is a strong taboo against participation in religious or other auspicious ceremonies during menstruation. Most married women, especially after the birth of offspring, do undertake cooking, but maintain a distance from the domestic altar. Being or having become outside *(eeche aguvudu, horagaguvudu)* is a common way of describing this state, which ends with a ritual bath. This act of pouring water (over oneself)—*neeru haki-kolluvudu*—renders a woman pure again, facilitating her (re-)entry inside (olage).

Another significant conceptual scheme surrounding menstruation, apart from the framework of pollution, is that of the girl being 'ready' for marriage and childbearing. Food practices observed with reference to a pubertal and immediate post-pubertal girl reinforce the latter perception, though care is taken at the same time to restrain her emergent sexuality. The rationale seems to be to provide food which will nourish the body, but also to avoid spicy or cold foods which could over-stimulate the senses. For instance, during this period, the girl is fed with bland rice and ghee, which is explained almost without exception by the phrase 'it is given to strengthen the waist' (*sontakke bala barali*). This strength is seen as important for an easy pregnancy and delivery. As repeatedly emphasized: 'The mother is the creeper. If that itself gives way, how can it bear fruit?' Food regulations thus serve to regulate the menstrual cycle and strengthen the reproductive organs, underlining the dominant perception of the female body as oriented to reproduction.[4]

Of Honour and Shame

The foregoing context, it should be evident, embodies certain ideas regarding the nature of the female, which are either privileged or regulated through specific practices. There is the conception of the fertile woman as potent, possibly polluting and potentially dangerous, which is sought to be channelized in both ritual (purificatory bath) and practical (marriage, motherhood) terms. The puberty ritual itself, by marking and proclaiming adult female status, also defines and prescribes the nature of her domestic role and the social spaces henceforth open to her. The experience of menstruation, the first time and every time, is a daily enactment of spatial demands. It entails not just physical segregation but also a temporary withdrawal from involvement in social activities. In 'becoming outside' (that is, by menstruating), the woman forfeits her claim over her conventional spaces, especially her participation in the inner, domestic realm.

A further form taken by ideas of 'appropriate spaces' for a girl involves the modes in which her physical mobility is subject to scrutiny and regulation (for instance, discontinuation of high-school education which would entail the girl going to the next village, a few kilometres away). Statements like, 'Isn't honour important above all else? What's the use of learning in the absence of that?' or 'What good will learning do to a girl?' further reinforce the implicit assumption that women's domestic chores need a different kind of capability. Combining with these tacit rules of space which stake out boundaries of the olage are reminders about 'proper' behaviour and speech for girls. Dress, for instance, is a powerful code, indicative of a girl's (and her family's) perception of her role and capabilities, as well as the entire community's recognition of this. Daily cautions of female propriety take the form of reminders to 'wear a sari',[5] or exhortations to sit decorously with knees together ('Are we men, to sit with our legs spread?'). These notions of female propriety and morality, ranging from posture and movement to dress and speech, are easily assumed and reiterated, while transgressions are commented upon and strongly ridiculed.

The transition from girlhood to womanhood is supported by implicit instructions from the wider community as well as the mother. The girl is chided about needlessly going out of the house and, when inevitable, she is told to go about her work with a bent head. Excessive movement outside the home, a readiness to laugh or smile, untied hair or clothes— all are read as signs of loose character. Especially after puberty, she is enjoined to speak less and softly, walk decorously and exercise restraint in dealing with men.[6] With this is entailed the overt teaching of household tasks, especially cooking.

Indeed, a girl's social (and physical) spaces as well as work activities become increasingly well-defined after her coming-of-age. The time lag between puberty and marriage is seen as a risky one, exposing both the girl and her family to scandalous gossip which may often be unwarranted. Yet, in spite of this, there is no direct talk about sexual morality, except in the form of admonishings, beatings, or even gossip. As Sharadamma, a Kuruba woman from Shivapura, put it:

> They learn a great deal from what is happening around them. Any slackness in the behaviour of girls will naturally be commented on by people. And they get to hear all this talk...that her family's name was spoiled, that she went with somebody and is now with child, that she is still unmarried, and all that. Then they become careful, but only after they understand.

However, a widespread assumption is that in spite of knowing the implications of all this, a girl's sexual feelings may be so strong that she could throw restraint to the winds and give in to a man's advances. Her alleged inability to exercise restraint is seen as necessitating external controls on female sexuality, especially in the form of an early marriage and motherhood. Shivalingamma of Hulaguppe shrewdly observed:

> If a girl who has come of age is often seen visiting other houses, people will gossip and doubt her character. But if a woman has a baby on her hip, no such problem arises. In a way, the child protects the mother's honour.

There is a general consensus that early marriages are best ('there will be no scope for fooling around') in order to control and channelize the active/emergent female sexuality, and also to guard and foster female virtue. A common refrain is that a grown girl should not be kept in her parent's house. Not only is delay in her marriage seen as unnatural, in the sense of disregarding natural urges, it is also considered damaging to the family's reputation. This finds expression in a number of metaphors and cultural idioms: 'You cannot set a match to dry straw and then ask why there is a fire', or more commonly, 'a girl should be careful, she is like an earthen pot; once dirtied, it cannot be used again'. Muniamma, a young Budabudike woman whose husband had recently married again, even urged:

One should not keep daughters in the house. It does not matter whether the man is lame or ugly. Just finish the marriage and send her off. Otherwise people will take advantage of her situation, assume she is of loose character. Don't I know all this...? Everything will be fine if she is married within six months or a year of maturing. 'He has no looks, he is dark, he is too fair, he is short'...no, let there be no excuses.

Marriage and Fertility

Female sexuality thus gains legitimacy only in the context of marriage and childbearing. Pre-marital sexuality is therefore theoretically taboo and procreation is culturally spaced so as to coincide with the married state. Marriage is a critical turning point for a girl, allowing her to translate into reality latent reproductive capacities and thereby realize her claim to womanhood in the eyes of the community.

Finding a suitable groom for a daughter who has come of age is, as we have already noted, a major preoccupation of the girl's family, relatives and friends. Caste endogamy is strictly followed and most groups practice certain preferential forms of marriage such as cross-cousin or uncle–niece marriages. But increasingly, there has also been a demand for dowry in the village cluster, not only among the upper castes, but also widely prevalent among agricultural castes and landless labourers. Yet, while the practice itself is widespread, there is variability in terms of the size of the dowry offered and even of the items that exchange hands.[7] Very often, an inability to meet dowry demands results in marriage negotiations breaking down, leading to delays in a girl's marriage.

The pressing obligation on the part of parents to marry off a girl underlines the high cultural value attached to wifehood, a sentiment also expressed in the positive ritual importance given to a *muttaide* or *sowbhagyavati*. Often, a woman's space is inseparable from her understanding and acceptance of what is culturally defined as her dharma, including her duties as a daughter, wife and mother. Morality, chastity, fidelity to husband, restraint, maintaining family honour—all these are intimately bound up in this perception. While these ideals—especially a wife's dharma to listen to her husband—are frequently reiterated in daily life, it is necessary to be open to possible disjunctions as well. At times, the value attached to *pativratya* could be used strategically by women, as when a woman publicly acknowledged as a 'virtuous wife' has greater freedom of movement since aspersions would not be cast on her motives. In some cases it is used to question another woman's morality and condemn her allegedly loose conduct, by mocking: 'If she is such a great *pativrate*, she shouldn't have come out of the house at all'.

If early marriages are seen as one way of placing controls on female sexuality, the effort does not end there. Even after marriage, a woman's behaviour is believed to be in need of constant supervision, as suggested by the refusal to allow a wife to visit her natal home too often.[8] But more commonly, restraints are imposed on a woman's movements and activities within the village by reiterating that her primary responsibility lies towards her husband and family. Often, her conversations with neighbours or movements outside the house would draw acerbic remarks from her husband or elders, warning her of such 'excessive' behaviour and reminding her of her place inside the home. The stress on chastity, morality and decorous behaviour then is not just a pre-marital requirement, but operates both before and after marriage, involving both personal and familial honour.

The linkages sought to be established between the female body, sexuality and morality, and the peculiarities of the honour code for women can be encountered suggestively in the differential use—across three diverse contexts—of the phrase 'women should bind their stomachs' (*hengasaru hotte kattabeku*). In one instance, a young Dalit widow used the phrase while referring to the economic difficulties and social accusations she faced while bringing up her children: 'I brought up my sons by binding my stomach; I often thought of ending my life, but that would also have caused tongues to wag...they would say I had affairs...'. Binding the stomach could here connote not merely controlling hunger but also placing restraints on the body's needs and desires. In another context, a Vokkaliga woman uttered the phrase 'women should bind their stomachs' as a moral injunction, commenting on a young Budabudike mother. According to her, while giving birth is an unavoidable and natural process, it draws attention to the woman's body and sexuality, which should consequently be properly bound or contained. A third context brings out further the moral overtones of the phrase, as referring not merely to the actual physical body and its needs but also to its perceptions in cultural terms. Speaking of the need to be careful in one's social interactions and conduct with men, a Lingayat woman remarked: 'Even if you go about with one hand on your stomach, the allegations will not disappear'.

The accretion of meanings exhibited by references to the stomach—or binding the body in different contexts—seem to signify an imperative to control and spatialize. Marriage, it must be reiterated, is central. However, although monogamous unions are largely the rule, instances of extra-marital liaisons, bigamy and even desertion of the wife are not completely absent in the village cluster. Though formal divorce is too alien a concept, 'incompatibility' often takes the form of outright desertion on the part of the husband. Not having children or sons is always cited as a major reason for desertion, though excuses like bad housekeeping are also used against the woman to help the man enter into another marriage.[9]

Quite different from such bigamous unions is the operation of illicit or extra-marital liaisons, where the woman is rarely from the same caste or village. She is usually from a lower caste and not expected to flaunt the liaison with a man from the dominant caste. In cases where offspring result from such affairs, they may make no legal claim to the man's property; sons of the legitimate marriage are seen as the rightful heirs. This difference in societal perception is brought out in a woman's retort to her husband: 'Am I the one you tied a *tali* on, or the one you keep?' (*nanenu kattikondiruvavalo, ittu kondiruvavalo?*)—made in the course of claiming her rights as a wife and reminding the man of his obligations as a father and husband. On their part, most women adopted a strongly moralistic tone, disapproving of extra-marital relations: 'Can one live happily by cheating the husband?'. A more down-to-earth view finds expression thus:

> *Can a woman afford to behave like that? What will happen to the family then, what will become of the children? Can I say I will leave my husband because he is a drunkard? Its like complaining that my nose is not straight, or that my mouth is crooked. Can it be ever corrected? ...One must continue to live, with the little one has.*

In addition to spatializing and normativizing issues of sexuality, restraint and control, the situation of marriage also provides the context within which female fertility is legitimated. This entails the representing of motherhood (*tayitana*) as a highly valued state, with infertility (*banjetana*)—or the absence of reproduction—being condemned as the failure of the feminine 'project'. Accompanying the positive evaluation of motherhood are also certain cultural ideas regarding conception and procreation. Irrespective of acknowledging the contributions of the man and the woman in conception, a child is also believed to be a gift from god (*devaru kottaddu, devara anugraha*). This nexus between divinity and the human is a belief that is dramatized, for instance, in the vows taken by a childless woman either to atone for past acts or to propitiate/ deflect an evil eye believed to be the cause of her childless state.[10] The metaphor of seed and field (*bija* and *kshetra*)—where seed represents the male contribution and field the female—is another dominant cultural notion regarding conception. By implication and context, this metaphor suggests that it is the quality of the seed which is of primary importance in determining the nature and identity of the offspring.

Dube (1986) provides a seminal analysis of the ideology of seed and earth, demonstrating how the restraints placed on women's sexuality tie in neatly with prevalent systems of kinship and patrilineal descent. She also goes on to argue that though popular notions of reproduction do recognize a close bond between mother and child, the bonding often becomes a context wherein maternal obligations and sacrifice are under-lined, rather than one where the question of rights over offspring is addressed. While appearing to accord a place to the female role in con-ception, the emphasis on maternal duty in such a popular discourse sets limits to the mother–child bond and, consequently, offsets any possibility of a challenge to structural (read, patrilineal) arrangements. Within popular perception in Aaladapalya, this close reciprocity between mother and child is articulated in diverse ways, but tends to cluster around the mother's role in nourishing the foetus. The child's temperament, for instance, is believed to be directly influenced by the mother's thoughts and deeds during pregnancy. Again, a pregnant woman's cravings for specific foods is seen as emanating from the child inside the womb. Even after the birth of the child, this relationship continues to be expressed in bodily terms, as when a number of dietary taboos come to be enforced on the mother. Since different kinds of food produce different flows from the body, the manner and type of food eaten by the mother is seen as affecting the quality of her milk, which, in turn, is held to contribute to both the health and character of the child.[11] This close link is further emphasized in observations suggesting that the child's body would ache if the mother worked too hard. Sexual relations with the husband are not resumed until three to four months or more after birth, since sex is thought to have an adverse effect on the quality of milk, thereby causing digestive disorders in the child. All these ideas amount to an implicit recognition of the link between mother and child, which could be placed as a counter-point to the dominant understanding of conception that emphasizes the substantial contribution of the father (see also Misri, 1985). Likewise, although the overall perceptions and descriptions of sexuality and pro-creation are in terms of 'duty' or the dharma of a husband and wife, where *kama* as an aspect of marriage is played down, a sensitivity to everyday contexts would reveal contrary perceptions in the popular consciousness. Though none of these countervailing perspectives get translated into structural terms, for instance, into changes in forms of kinship or property transmission, their significance in affording a traversal of the everyday cannot be denied. The body is not simply a dimension of experience, but rather the form in terms of which experience is organized.

A Matrix of Sexualization

As the foregoing sections suggest, references to the feminine rarely take a direct form, being generally mediated through cultural idioms, values and practices. These ideas, often spatially orchestrated, also bear upon a distinct and circumscribed context, namely, the domestic domain. Even women's perceptions of themselves and their bodies are routed through these matrices and spaces, a fact which makes it necessary to see how the physical or biological substratum gets housed within cultural notions of the feminine.

By way of addressing this issue, one may begin by making a broad analytical distinction between two aspects of femininity—as biology and as socio-spatial construction, though both are inextricably linked in reality. The former marks a reference to those biological aspects which define a woman, such as her procreative capabilities. Though there is a tendency to reduce this to the natural order alone, it may be seen that both in experience and perception, these aspects are culturally mediated and represented, as when reproduction gets channelized through marriage and the legitimacy of motherhood. Femininity as a socio-spatial construction refers to the elaboration of a moral order based on an acknowledgement of the biological substratum. However, this does not automatically confirm or re-state the natural order. It is based on the natural order only to the extent that it builds on a recognition of, for instance, physical changes in the female body, sexual urges, and so on. The dialogic relation between the natural and the socio-spatial appears to be a transcendent one, where the biological substratum is not denied; rather, the effort is to control and channelize (or transcend) the natural by incorporating it into cultural idioms and practices. Thus, though femininity is based on a recognition of the biological substratum, the moral order seeks to spatialize and control biological facts, in the process achieving a transcendence whereby femininity becomes a product of socio-spatial emendations rather than strictly biology.

It is important to reiterate that the biological is not denied and is very much a part of the constitution of femininity as a socio-spatial and sexual fact. This is clearly evident in the cultural perceptions of pubertal girls and the widespread insistence on an early marriage for girls. The crux of the matter is not a non-recognition of biology, but the modes in which this reality is sought to be (en)countered and deflected in socio-spatial terms. Demonstrating the mutual exclusivity of the 'natural' and the 'cultural' therefore is both misleading and distortive of the contextual realities. The challenge, rather, is to grasp the inter-relations between the two, a task that is often complicated since the biological and moral orders are not uniformly integrated into the discursive consciousness of the community. The biological underpinnings are recognized but rarely spoken about, while the moral order seeks to interpret and represent the biological in distinctive terms. Any understanding of the construction of femininity will, accordingly, has to take into account the inter-relations between the biological and the socio-spatial. This would also be facilitative of a movement beyond the formative distinction underscoring gender studies—that between 'biological' sex and 'social' gender. Indeed, we come to recognize that gender is implicated within socio-cultural practices regulating sexed bodies.

The point of the discussion so far has been more to show how female bodies, at every step, are spaced, ensconced in layers of meaning and belonging than to draw attention to socialization practices or to the different stages of the female life-cycle. Such a process of female embodiment is facilitated through their insertion into what may be termed a *matrix of sexualization*. The matrix specifies certain codes of moral conduct within the community and is often responsible for the active espousal of conceptions of the feminine. As elaborated in this chapter, it contains injunctions regarding shame and honour, marriage and motherhood, and working towards regulating the movement of women, their activities, dress and even speech. In linking cultural notions of femininity with an underlying biological substratum, it serves to orient female dispositions as well as mediate in women's bodily practices within the community. It is also clear that this matrix is engaged in drawing the spatial parameters within which women's lives are defined. As we have seen, the rules and experiences of menstruation and the legitimation of fertility through marriage, as indeed the attendant ideas of honour and shame, all serve to carve out the contours of an inner, domestic, female domain. Focusing on such a matrix and its operative dimensions can provide a passage beyond the deadends in gender analyses that we have noted. More explicitly, it enables us to infuse the constructivist turn in gender analysis with a materiality deriving from a direct focus on bodies. Along this line of appraisal, of course, several other questions crop up, such as: Where does the female subject stand in this altered constructivist account? How does she respond to the socio-spatial and discursive matrix of sexualization that envelops her? It will clearly not do to suggest that the matrix imposes itself on women, or to speak of the female actor as a translucent self, responding autonomously to a body of impositions. Rather, it is necessary to conceive of a female bodily subject who is both an agent (a source of action) and a subject (subjected to a set of rules or laws preceding her) (Meijer and Prins, 1998). The woman, in 'acting', not only confronts a set of normative injunctions and ideals, she is also constrained to shape herself in relation to them—and not only in order to en-gender herself. The next two chapters will seek to further complicate the terms of this sexualization/spatialization, even as Chapter Six addresses the specific 'burden' of agency that my narrative seems to implicate.

Notes

1. The observation regarding the body as a central metaphor for society, made by Das (1988a) with reference to analyses of caste and ritual, can be extended even more strongly in a study of gender and space.

2. Bourdieu (1990a: 77) has termed this system of shared values the habitus: '(T)he habitus, as a system of dispositions to a certain practice, is an objective basis for regular modes of behaviour, and thus for the regularity of modes of practice, and if practices can be predicted, this is because the effect of the habitus is that agents who are equipped with it will behave in a certain way in certain circumstances'. However, I shall refrain from weaving this notion into the body of my narrative. To be sure, Bourdieu does show the way, but I suspect the notion is over-theorized. As will become evident, I prefer to speak of a matrix of sexualization as a precipitate for the habitus and its effects.

3. The grass hut is built by the girl's mother's brother and comprises of leaves from three trees—*atti, aala* and *bevu*. After she is bathed and admitted into the house on the third day, the shelter is removed and destroyed under cover of darkness. On the sixteenth day, the *osage* ceremony (or *rtu-shanti*) is held. This ceremony highlights the fertility of the girl, as underlined in the major rite of symbolically 'filling the womb' (*madilu tumbisuvudu*).

4. I shall return to this point more actively in the penultimate section.

5. The sari suggests a complete semiotic, its wearers being identified as embodiments of morality. Typically, ideas of shame and honour are reflected in the changing dress of a girl: who grows out of a frock in early childhood to an ankle-length longskirt in adolescence. Often a *davani* is added to the longskirt (symbolically covering the bosom) and signals a clear pre-marital code, either anticipating or marking her coming-of-age.

6. In the village, loud talk, laughter and a rough appearance immediately provokes identification with the image of Mulkatamma, a single village goddess. Likewise,

the opposite characteristics in a girl/woman earns the comment that she is the Lakshmi of the house. I shall be invoking the space of gods and goddesses in Chapter Five.

7. Dowry demands are not always direct, nor do they always take the form of hard cash; it typically includes jewellery, clothes and/or other household articles. Sometimes, the bride's family is asked to pay off a debt incurred by the groom's family; at other times, the boy's side indirectly indicates that they spent a specific sum on their own daughter's marriage, thereby hinting that they expected that amount to be made good at their son's wedding. Even within impoverished families, the practice is widespread. Mothers of marriageable daughters complained that even a man with nothing but a plough to his name would ask for a sizeable dowry.

8. Some women feel that their husbands and/or in-laws would never send them to their natal place out of fear that she would take up with another man there.

9. According to some, the 'permission' or 'agreement' of the first wife to the new alliance is necessary. If unwilling, she could seek justice from the village panchayat. Remarriage, especially after his wife's death, is entirely conceivable for a widower, though it is unheard of for a widow among the higher castes. The lower castes, however, do allow widow marriages, known as *kudike*. This practice is also extended to women who have been deserted by their husbands and, in particular, to deserted women childless from the first marriage. The kudike practice involves the marriage of such a woman to an older man, who is himself either a widower, separated, or past the usual age of marriage. There are no special rites or ceremonies performed, though a small kin group is invited for a meal and the couple may exchange garlands. It is pointed out that this woman cannot wear all the insignia of a sowbhagyavati. She also cannot participate fully in any auspicious or ritual occasion.

10. The most common ritual response to childlessness is to take on a vow (*harake*). Sometimes, inability to conceive is attributed to *nagaradosha* (literally snake defect), which needs to be corrected. *Nagara pratishte*, a specific ritual practice, is the most commonly accepted mode of correcting this.

11. At times, the food eaten by the mother is said to cause indigestion in the child. Specific foods such as plantains are alleged to cause wheezing in the child, whereas consuming the neck of a chicken is believed to make the child's neck firmer.

Five

Carving Ritual Spaces

Studies of religion at the village level had, conventionally, sought to differentiate between a Sanskritic tradition and a non-Sanskritic or local one. Increasingly, however, what is becoming obvious are the porous boundaries between the two traditions, a recognition that has made our renderings of religious practices that much more complex. In reality, religious practices rarely, if ever, inhabit water-tight compartments; more usually, they assume meaning and power within definite historical and situational contexts where the two traditions (namely, the Sanskritic and the non-Sanskritic) are either in interaction or in tension with each other. Therefore, instead of deploying this over-worked differentiation, I shall try to work round it by examining a range and multiplicity of ritual practices, focusing in particular on how women inscribe themselves in, and through, such practices.

To be sure, ritual practices may be approached in different ways: in terms of their degree of formalization, their orientation (whether directed towards personal ends or common welfare), and the nature of involvement (whether individual, family, caste or entire village); in terms of their classical or indigenous referents; and in terms of their temporal rhythms (whether fixed/cyclical or more flexible, personal forms). I seek to combine and relate two aspects in this discussion—one, the nature of involvement and participation in the ritual and two, its fixed or variable character—and in doing so distinguish between three kinds of ritual practices. First, a ritual may include a woman and her household and take the form of a *harake* (vow) or *vrata* (rigorous worship), both of which are non-cyclical and relatively short-lived in nature. Second, when a ritual is repeated regularly by a woman, household or larger group, it assumes a certain fixity which is conveyed by the term *habba* or festival. Several annual cyclical festivals like *Sankranti* and *Ugadi* are commonly celebrated by households, whereas women of certain castes celebrate the quasi-ritualistic *varamahalakshmi* and *gandana puje*. There is also an annual *uru habba* involving the entire village and dedicated to the village guardian deities. Third, there is the annual *jatre* or *utsava*, dedicated in Aaladapalya to the goddess Kunigalamma as benefactor of an entire region.

At each of these junctures, the ritual practices are highly stylized and symbolic, and though followed meticulously by the participants, their complex meanings are either not verbalized or are differentially articulated across them. My task in this chapter is not of course to plumb these depths of popular Hinduism, but rather to examine how certain ritual practices lend themselves to a delineation of social and gendered spaces as well. Apart from specifying what the contours of such spaces are, the focus will also be on the manner in which women carve out ritual spaces for themselves. The attempt is to examine the inscriptions of the feminine at every layer, to see how it translates into a certain perception about women and their relations to (and negotiations of) space.

Taking on Harake, Observing Vrata

The religious conjuncture animated by harake and vrata is unique in that these are predominantly women-centred acts. Harake is a ritual observance that involves the woman taking on a conditional vow, one that links her closely with the divine. 'To take on' harake (*harake hottukolluvudu*) is to seek a deity's intervention in tackling problems, whether it be illness, misfortunes, delayed marriage, childlessness or even unemployment. It also entails a promise of repayment to god for favours bestowed, either in the form of special worship and offerings to the deity, gifts of money, animal sacrifices, or even by taking on physical trials as a token of one's gratitude. Harake, accordingly, is not to be understood as a mere seeking of blessings but more as a pact between the person and the divine, a pact that must be honoured. 'Taking on' harake is closely related to obstacles (real or otherwise) in the course of progression through one's life-cycle (for instance, in marriage or childbirth). Often the fulfilment of a harake vow may coincide with, or sometimes form part of, a particular life-cycle ritual itself (for example, offering the hair of an infant to god, or the mandatory visit of a newly married couple to the temple where a harake has been taken); yet, it is sometimes necessary to distinguish between the two. For one, life-cycle rituals (say a marriage rite) are by definition 'one-shot' events, quite unlike the cyclical, repeti-tive quality of such configurations as harake.

Unlike the harake, vrata observances are better described as obligatory rites, often seen as a duty by women within the domestic sphere. Vrata observance or performing vrata (*vrata acharisuvudu*) is aimed at invoking divine beneficience in order to ensure the well-being of one's husband and children and, by implication, one's own location as wife and mother. Sometimes, what begins as a conditional vow (harake) can, upon fulfil-ment, get transformed gradually into an obligatory rite (vrata). This con-tinuity was obvious in the case of Seethamma and her daughter from the Brahmin community. In the former's words:

> I had taken a vow (harake) that if Revathi's marriage is fixed, she would undertake a vrata to preserve her sowmangalya. I myself had observed the Mangala Gowri vrata, for the well-being of all in the family. In spite of this, the auspicious moment has not arrived. Who knows what is written on her forehead? We have been told that the next few years are not very good for her. Maybe she should begin a vrata right now so that she gets a good husband later.

However, it would be misleading to gloss harake as a 'short-term vow' (Hanchett, 1988: 301), contrasting it, in effect, with vrata as a more pro-longed vow, for their very natures are distinct.

'To take on' harake is, at the very point of initiation itself, different from the 'observance/performance' of a vrata. The latter is elaborately ritualistic, with definite procedures to be followed in its performance, a feature absent in a harake, where the pact entered into can vary depending upon the perceptions, capacities and inclinations of the person taking on the vow. Demanding a pledge of devotion and commitment on the part of the woman undertaking it, vrata observance also establishes a personal and prolonged relationship with a deity. Besides, vrata is almost entirely myth-based, its observance/performance including a narration of legen-dary stories tracing its emergence, the widespread devotion it attracted, and guidelines as to how it is to be performed. The 'transformation' of harake into vrata, therefore, is not necessarily an inevitability, but the result of the devotional attachments formed by particular women whose religious practices then begin to take on a structured character as precise points in space.

By and large, unlike harake which is more prevalent among lower castes, vratas constitute a major chunk of the religious observances of upper-caste women. A large majority of the Brahmin and Lingayat women in Aaladapalya undertook such vratas, prominent among these being the *Mangala Gowri vrata*. While the tradition of observing this vrata suggests that it is optional, it is however rendered obligatory in effect by the insistence that it is one of the duties of a *pativrate*. Directed towards preserving the well-being of the husband, this vrata, by implication, is concerned with protecting the state of a sowbhagyavati. It is also known to bestow long life on a son and to destroy possibilities of a daughter's widowhood. Essentially, Gowri is worshipped as a 'married woman goddess' and is perceived as the protector and benefactor of married women.

It is important to mention that apart from the objectives towards which vratas are directed, the very course of their observance—whether in terms of fasting, observing *madi* and other rules of pollution, or undertaking related acts of worship—introduces changes in the daily routines of life and perceptions of the people.

Festival Time

It is also possible for a vrata to assume, over time, the status of habba (festival). Though this is specific to certain castes, or even to certain families within a caste, the regular observance of a ritual over several generations of a lineage, or by several families of a sub-caste, confers on it a certain fixity in ritual time and space so as to be designated habba. For instance, unmarried girls as well as married women from the higher castes in the village regularly celebrate such vrata-based festivals as varamahalakshmi and gandana puje. The latter is for the protection of the wifely status and is celebrated as a mark of gratitude to Shiva for having granted *sowmangalya* to women. The former is a similar rite, oriented to ensuring an increase in wealth, grain and male offspring. Its accompanying myth begins with a recommendation that this is a ritual which will bring well-being and prosperity to all who perform it, followed by shorter stories demonstrating its efficacy in particular cases. Besides reciting these mythical stories, women celebrating this festival also gain reassurance by hearing of other women who have supposedly benefitted from it, seeking thereby to relate it to the alleviation of problems or suffering in their own lives, brought on by poverty, childlessness or marital/familial discord. As is apparent, women act as the conduits through which flows of well-being are invoked and ushered in. These flows, based on the 'strength' of the woman's devotion, bear essentially upon the space of the domestic, the household in particular: *kutumbakke ayussu, arogya, sukha, sampattu sigali* ('may all in my household be blessed with health, happiness, wealth and long life'). It is in this sense that women (across castes) may be spoken of as bearers of auspiciousness within the family.

In addition to the festivals celebrated by women of the upper and intermediate castes for their personal and familial well-being, there are other major festivals involving the entire community. On such occasions, the role played by women extends to the domestic side of the ritual worship, in a way demonstrating their supremacy within this realm. Besides, tacked on to the ritual significance of a festival is its suggestion of a re-generation of the community's energies, partly embodied for women in periodic cleansing of the house and its environs.[1] In festivals such as the harvest festival, which marks the productive rhythms of the entire agricultural community, the role of women appears to be as mediators of purity as well as bearers of auspiciousness. This is true both on the occasion of a classical festival (like Ugadi, which is celebrated with minor variations across South India) and during a much more local festival like the annual village festival (uru habba). Let me traverse the ground of the latter as encountered in Aaladapalya.

As the very term uru habba suggests, this festival is not extensively celebrated over an entire region but is specific to a particular village or, more appropriately, uru. Its local flavour derives from the fact that the festival is in honour of the major guardian deity or deities of a given village, in this case Basava and Veerabhadra (largely Sanskritic deities) and Mulkatamma (a non-Sanskritic deity) of Doddaaluru. While the former two deities are the central focus of the uru habba for all except the lowest caste group, this is followed by a brief propitiation of Mulka-tamma by the Dalits on the day following the festival. Though there is no fixed date for the holding of this festival, it is usually held nine to ten days after Ugadi. As one elderly resident of Doddaaluru put it: 'Ours is God Rudra (implying anger, impatience). He won't wait long.' Every year, the elders and prominent men of the village meet to decide when it should be held and when to begin preparations for it; their decision is then announced throughout the village.

The festival, as already mentioned, centres around three guardian deities—Basava, Veerabhadra and Mulkatamma. The former two are described as upper-caste or Lingayat gods whereas the latter is worshipped by the Dalits. The Basava temple, which marks one of the outer limits of Doddaaluru (to the west), is relatively small and there has been some talk of renovating it. The outer portion comprises a raised verandah which opens inward into a square enclosure. Straight ahead is the *garbhagudi*, which houses the Basava (Nandi). The Veerabhadra temple is towards the eastern edges of the village, rather removed from the habitat. It stands on the banks of a natural tank and is extremely run-down. (Veerabhadra is often cited as the *nele devaru*—the founding god of the village.) Mulkatamma guards the southern edges of the village, her place being simply marked by stones with no elaborate shrine or temple.

The actual activity and excitement begins on the night prior to the festival day. The *jangama* (a sanyasi, not the priest) offers prayers at the Veerabhadra temple, after which he dons a *hulivesha* (tiger's dress) and performs (*kunita*) throughout the night, moving from house to house through the entire village. Against his body hangs a wooden plank, said to represent the god Veerabhadra. During the kunita, he also relates episodes from the Shiva–Parvati legend. As he dances, four other men, carrying the Basava, are in attendance. Burning cudgels, lanterns and small bulbs light up the performance and, mingling with the drumming, creates a deafening and riveting scene. As the jangama pauses before each house, the women of the household offer *puja*, breaking coconuts to pacify the god. Those who are unable to join the next day's worship offer *arati* at night itself. This is especially true of the very poor, or if there is nobody at home to carry the arati, and/or if they belong to the Dalit community. While the deities stop before each house throughout the main village, this changes at the residential cluster of the Dalits. The gods wait at one place, while the women and girls come to offer arati.

The next morning, when the crier boy announces the Basava arati, there is again a flurry of activity in the houses carrying arati to the Basava temple. The arati carriers are chiefly young girls or even older women— widows neither carry nor offer arati—and they move in a procession to the Basava temple, where a *mangalarati* is held.

The focus of attention then shifts to the Veerabhadra temple, where the same arati is carried by the women and girls. Devotional songs or a *harikathe* recital is broadcast over loudspeakers. Preparations for a meal are also going on. The *konda*—a bed of wood and live coals—is also prepared, the crossing of which is one part of the uru habba. (It is noteworthy that most of the organizational details are left to the menfolk while the prime responsibility of the women seems to be to adequately represent the household in the worship, propitiate the deities and seek their blessings.) The konda is first crossed by the jangama after a ritual dip in the tank on his way to the temple. Women who have taken a harake before the god (especially seeking cure for ill-health) also cross the konda upon fulfilment. It was also observed that it was mainly non-upper caste, that is, non-Lingayat women, who crossed it, perhaps an indication that harake observance in this form is more widespread among the lower castes. Arati is then offered by them at the temple. Though the Dalits come near the site of the konda, they do not offer arati at the temple. Their non-participation (in relative terms) in the uru habba of Basava and Veerabhadra was explained by both lower-caste and upper-caste informants by stressing that their gods were different.

The festival of Mulkatamma, muted in contrast to the pomp of the other, is held the next day. The primary explanation offered for this was that, being an angry goddess, she demanded an animal sacrifice, and thus could not be mixed up with the 'vegetarian gods'. Mulkatamma has no procession throughout the village. A priest (*pujarappa*) from among the Dalits carries the representation of the deity on his head and dances, while families offer arati. Some families from the interior of the village also send the children (believed to be immune to pollution effects) with an arati for this goddess as well. The 'shrine' of Mulkatamma, with no distinctive anthropomorphic image, is at the southern village boundary, a direction usually linked with death and inauspiciousness; the animal sacrifice conducted here is intended to appease the goddess and to renew her energies in the battle with evil forces.

While Basava, Veerabhadra and Mulkatamma are all perceived as the guardian deities of Doddaaluru, guarding its boundaries against the infiltration of evil, the significance they commanded often varied. Several myths and legends regarding the deities reflect this variation. For instance, some indication of the transformation and incorporation of pre- or non-Sanskritic deities like Mulkatamma and Maramma into the Sanskritic corpus was provided by a Lingayat woman narrating the legend of Veerabhadra. The story recounted by her cites the deep humiliation experienced by Parvati as a result of her father's actions during a *yajna*. So intense and unbearable was her anguish that she jumped into the yajna fire.

In Kailasa, Shiva immediately sensed that something had gone wrong. He worked himself up into a frenzy and from the drops of sweat that he wiped off his forehead and cast aside, Veerabhadra emerged. Veerabhadra's nature was as angry and rage-filled as the emotion which created him. He rushed in his fury to attack Shiva, who created Bhadrakali to divert and pacify Veerabhadra's angry spirits. Controlling his anger, Veerabhadra asked who illtreated his mother, and Shiva explained the situation to him. Without wasting time, Veerabhadra set

off to annihilate Daksha Brahma, who had humiliated Parvati so gross-
ly. Veerabhadra's impulse was so strong that whatever he encountered
on the way was mowed down relentlessly. A trail of blood and destruc-
tion was created in the wake of his movements. The various ammas,
like Mulkatamma, Maramma, Chowdamma and Kunigalamma were
created by the gods to 'lick up' the bloody mess left behind during
Veerabhadra's mission....

Studies have pointed out that village deities are predominantly female
(Erndl, 1993; Fuller, 1992) and local in their origin and history. Yet, in
terms of symbolic and expressive practices in the field, there is an attempt
to encompass these goddesses within Sanskritic male principles, though
at the same time taking care to propitiate the heat and danger deriving
from their independent status. This recognition of ambivalence also
derives from the dual nature of such female goddesses, as we will see in
the case of Kunigalamma in the village cluster. Hitching indigenous
female goddesses to deities in the Sanskritic pantheon has perhaps worked
towards making the worship of the former more acceptable, especially
to the upper-caste groups, thereby erasing possible contradictions in the
worship of village deities. Nevertheless, perceptions such as those also
underscore a process of the hierarchical construction of relations between
the gods themselves. As we saw, Mulkatamma is described as an inferior
deity largely because she is among Veerabhadra's subordinates. Her
inferiority could also have to do with the fact that she is not a vegetarian
goddess, since the perception of deities as meat-eating or vegetarian often
becomes a metaphor for their low or high placement in the divinity (cf.
Dumont 1970: 27–30).[2]

The honouring of guardian deities within the village provides an interesting intersection of the territorial (the protection of village boundaries)
and the socio-spatial, since uru is itself lived as a cultural space defined
by one's caste and kin rather than as physical space alone. The uru habba
thus participates in a negotiation of what a community means by uru,
perhaps in the process also reaffirming the structural space of the community.

Jatre

This brings us to jatre as a form of ritual procedure, distinct from vrata,
harake and uru habba. The jatre pertains to a single local goddess,
Kunigalamma, and is distinct from other ritual forms in that it incorporates

within it several outlying villages (thirty-three in all), which are believed to be the domain of the goddess. Another distinctive feature of the jatre is that it takes place at a ritual site which is geographically removed from the actual habitation sites of the surrounding villages. The focus of ritual attention is neither the household/hearth nor the boundaries of a particular village, but the *dodda aladamara* (big banyan tree) and the shrine amidst the expanse covered by the tree.[3] For all the surrounding villages, the Banyan tree provides a means of representing themselves to others, a locus of ritual and spatial identity. As some women remarked: 'The Banyan tree is very central to our lives. It gives us a mark of identification (*gurutu*), so that when we speak to others, we can say we belong to this region'.

Kunigalamma is perceived as potentially destructive in nature and, when angered, can cause havoc in the villages. Her inherent malevolence is believed to require periodic deflecting and 'cooling' (*shanti maadi-sabeku*). She is worshipped every year, especially when there is an outbreak of an epidemic (either among humans or cattle), or fear of a drought, for these disasters are taken to be an indication of her 'angry' or 'heated' state. The worship is directed toward satisfying her heated appetite through animal sacrifice or by offering cooling foods, thereby paving the way for a flow of 'beneficient' energies. In many ways, she represents the mother goddess in the villages and is closely linked with the domains of fertility—human and natural.

The original Kunigalamma shrine (housing a stone icon) is at Marana-halli, a village about a mile away from the Banyan tree.[4] Ganapatihalli, a village formed when a section of families of Maranahalli splintered off, has a new temple for the goddess. The image here is made of tin and is the *utsava-murti* (the icon used for the jatre). Many people go to the older temple for worship, which is held every Monday. During the jatre, the icon of Kunigalamma starts from Ganapatihalli, goes to the older village, and only then is taken in procession to the Banyan tree, shrine of Muneshwara (Shiva).

The jatre is an annual phenomenon, usually held within a fortnight of the Ugadi festival. While the actual jatre is only for one day, the worship of the goddess continues over another fortnight, since she is taken to the thirty-three villages which comprise her domain. Bringing the goddess to individual villages depends largely on local initiative and contributions from the households in order to meet the expenses of the percussionists, bearers and priest. Once in the village, the deity is taken to households desiring to offer worship, where the goddess is propitiated and her protective support sought. The non-meat eating castes prepare an offering of curd rice (*yede*) and offer *mangalarati*. Some of the lower castes take arati on the day of the jatre itself and when the deity visits the village, animals are sacrificed. Killing of sheep is a must for the goddess and, according to informants, she will refuse to enter a village if there is no blood spilled. 'She wants blood...she is a goddess who sucks blood' is a common way of referring to her heated state.

This destructive and bloody nature of the goddess, which for the Vokkaligas, Kurubas, Dalits, Budabudikes and some others is a fearful but natural aspect of her, is intrinsically problematic for the higher castes like the Lingayats and Brahmins, though they 'dare' (*sic*) not desist from her worship. Accordingly, several rationalizations are offered. In the opinion of a Lingayat informant, animal sacrifice is not so much for the goddess herself but for her attendants (*sevakaru*) and her six other sisters. The fact that the sheep is sacrificed only after Kunigalamma's back is turned, probably a later accretion, is cited as proof of the above viewpoint. Another woman, who equated Maramma with Kunigalamma, insisted that these were all different *avatars* of Lakshmi. Annamma, Mulkatamma and Muttalamma were sisters and demonic manifestations of the usually benign Lakshmi. An old Brahmin priest, who cited the Kunigalamma festival as the main one in Hulaguppe, sought to trace this indigenous ritual practice to the Puranas and to relate it to the episode of Kamsavadhe.[5]

Such mythological elaborations suggest how a female deity, who is clearly external to and possibly prior to Brahmanism, is sought to be integrated into a more Sanskritic world-view. At the village level, this rationalization and incorporation works in such a way that the worship of very local female goddesses (concerned with local and practical issues) can co-exist with a belief in, and worship of, the higher gods (both male and female) of a pan-Indian, Sanskritic kind. The repurcussions this phenomenon can have on perceptions of femininity and female space will be considered shortly. But let me now return to the jatre.

On the day of her jatre, Kunigalamma is ritually transposed from her original embodiment as the immovable shrine icon into the utsava-murti, the image used in the procession. As the mother goddess, Kunigalamma belongs to a pervasive world-view which recognizes the centrality of the female principle and its regenerative capabilities. Yet, this recognition is a highly ambivalent one, since it also stresses the volatile and disruptive nature of the goddess. Harnessing this volatile force for the welfare of the community is the underlying motif of her worship. While

Kunigalamma in her angry form is associated with disease, death and a certain instability in social life, the course of the jatre suggests a gesturing towards renewal. The independent powers at her disposal are sought to be controlled, harnessed through her 'marriage' and by sacrificial propitiation, all of which are believed to culminate in a renewal of life. Thus, before the jatre begins, the utsava-murti (represented only by a head) is ritually 'married' to a wooden post symbolizing the deity Shiva. The goddess is then brought in procession to the site of the jatre—the clearing adjacent to the large Banyan tree—where there is a small temple of Muneshwara. The utsava-murti is carried by lower-caste men, whose energetic dancing is attributed to the infusion of the *shakti* of the goddess. The jatre draws large crowds and holds different kinds of meaning for the range of people who participate in it. For some, it is the splendour of Kunigalamma's powers (*mahime*) that attracts, while others are keen on offering ritual worship at the shrine and earn her blessings. Some seek to take a vow (harake), others come to acknowledge the fulfilment of a wish, usually through the payment of *muyyi* (a certain sum of money).[6]

Overall, the jatre provides a socio-cultural and religious space within which to reiterate and extend ties, both at the human and divine levels. Significantly, the extensive participation of women in the jatre serves to highlight societal and individual perceptions of their ritual roles and functions. Again, this participation does not extend to organizational matters; rather, women are held responsible for properly invoking the deity, offering worship, and in general mediating and sustaining relations between the human and the divine. The context of the jatre also transforms perceptions of spatiality and renders women highly visible. As an allowed ritual space, it appears to suspend, temporarily, the social demarcations of the inner (olage) and outer (horage) spheres. Perhaps this underlies the words of a woman who quoted a proverb: 'One should not attend a marriage accompanied by children, nor should one go to a jatre accompanied by the husband'. While this can make for varied interpretations, it is possible to posit that both marriage and jatre signify a certain acceptable space which are liminal, yet not dangerous, combining as they do the absence of dangers of an outer sphere and the extended security of the inner hearth.

Criss-crossings

Woven into the religious conjunctures of harake, vrata, habba and jatre considered above, are a number of intersections that connect individual ritual observances with wider community practices. Unlike harake and vrata which are both person-centred, the uru habba or village festival involves the worship of local deities in a ritual act that reinforces village boundaries and engages the entire village community in very specific ways. The jatre, which subsumes several outlying villages (urus) provides yet another, broader, ritual enclosure encompassing individual villages, and even enables a certain delineation of a 'region'. What I seek to suggest is that the meanings of the religious within the village are not entirely grasped if the underlying spatial dimensions go unrecognized. In other words, religious/ritual practices, besides engaging in transactions with the divine, also mobilize space in definitions of community. What is of added significance is that this delineation of sacred spaces conveys an understanding of gendered spaces as well.

How does this gendering of sacred and social spaces occur? To draw on the preceding discussion, we can see how harake and vrata (as well as uru habba) participate in retracing the parameters of an 'inside' and 'outside' as those primary social spaces which members identify with. As has already been pointed out, olage–horage are not necessarily fixed entities defined primarily in relation to each other, but also assume more fluid connotations in relation to a range of other entities, be it the household in relation to kingroup, caste, uru, or even the town/city. The widespread tendency to characterize the inside as a space appropriate for women turns on predominantly moral grounds, being hinged on such issues as female chastity, caste purity, family honour, and so on. Important for this discussion, however, is a quality firmly attached to this inside—namely, its inviolable and exclusive nature. Its inviolability reinforces its 'moral' overtones, delineating not only the space exclusive to women but also, in effect, drawing boundaries around the domestic, the conjugal household in particular.

Harake and vrata exemplify this conjunction. Their observance is women-specific, explicitly oriented towards ushering in a state of well-being for the familial/domestic realm as a whole. The fact that it is the married woman (sowbhagyavati/muttaide) who is largely responsible for protecting her people from misfortune and for invoking the flow of auspiciousness is evident even from the ritual role played by such women in the uru habba and jatre. Analogously, the ritual marriage of Kunigal-amma and Shiva during the jatre may be read as a gesturing towards renewal, whereby a potentially inauspicious figure—the goddess as someone who brings disease and death—is not only deflected but also sought to be transformed into an auspicious force through marriage. What comes across strongly in such contexts is the affirmation and celebration of conjugality as a continuing state and of marriage as inaugurating a flow of auspiciousness in the woman.[7] Not detached from pleas for prosperity, well-being and good health, it is this axis which characterizes the spacing of the religious within the community. Perhaps one could also posit that the domestic, far from being a subordinate realm that is encompassed by the religious, serves to actually ground the latter.[8] The stress on marriage and familial well-being in particular serves as a vehicle for 'this-wordly *bhakti*' (Erndl, 1993: 159) within the domestic realm, a devotion which is not so much spiritual but one which has definite links with demands in this world.

Another Spacing

The significance of the domestic space, even as a purely ideological frame, is apparent when we examine the orientation towards female deities as well. This is invariably approached as an alternation between the 'benevolent' and 'malevolent' aspects of deities.[9] The benevolent quality—as represented by the Mangala Gowri vrata among upper-caste women, or even the varamahalakshmi habba—is embodied by the Sanskritic spouse goddesses, that is, those married to gods in the Hindu pantheon (for instance, Gowri and Lakshmi), and emphasizes the harmonious and productive role of married women. These goddesses stand for wifely traits and their powers are manifest in conjunction with the spouse. They are emphatically benevolent, with their jurisdiction extending primarily to domestic concerns (see also Das, 1988b and Gatwood, 1985). Granting household prosperity, productivity, fertility, good health and offspring are some of their main functions. Their point of reference is the household (and not just the woman) as a symbol of fertility and domesticity. Emblematic of the malevolent dimension is the village goddess Kunigalamma, who is less concerned with the domestic domain per se than with protecting and augmenting the health and wealth of the entire village community (cf. also, Ganesh, 1990). She also stands out by being an autonomous goddess, whose shakti is sought to be channelized through a symbolic marriage to Shiva on the day of her jatre, as well as by propitiation through sacrifice.

Be it in terms of their manifestations or their forms of worship, these deities are not uniform, but reflect (as I would like to aver) a *near*-duality. In spite of addressing different social spaces, it is possible to suggest that the two kinds of goddesses share a commonality—in that they serve to regulate a range of biological and social reproductive activities in the community. Whether it be Gowri as protector of married women or Kunigalamma as benefactor of an entire region, both underline the regenerative powers attributed to female goddesses. The fact that there is no deep-seated schism between the two, both in perception and practice, is also evident from the co-existence of the worship of the single goddess along with the spouse goddesses in the village.

It may thus seem to appear that benevolence and malevolence are but two sides of a common female principle. The figure of Kunigalamma conveys these qualities quite aptly. Indeed, much of the existing discussions about Hindu female deities have also been structured around this contrast (see Erndl, 1993). Nevertheless, we need to ask whether such a structuring really captures the complexity that is at the heart of such ritual practices on the ground. Let me clarify this by drawing attention to the hierarchical principle of caste society, which, incidentally, can be rendered as a distinct spatialization and which also seems to underly the ordering and ranking of deities.

The goddess Kunigalamma herself may be approached as an embodiment of the hierarchical principle, albeit at multiple levels. This is obvious both from the events marking the jatre as well as from the rationalizations offered of specific practices. Her ritual marriage to Shiva, a 'high' Sanskritic deity, represents—on a theoretical plane—an encompassing or her subordination to one higher in the hierarchy. And yet, it is difficult to assert this conclusively, for it is not Kunigalamma herself, that is, her 'original' embodiment, but her transposed image (the utsava-murti, represented by a head) that is 'married' to a wooden post symbolizing Shiva. In other words, while we talk of her subordination to a higher principle, we must, at the same time, also concede that there are aspects of her ritual practice which suggest that the encompassment is incomplete; indeed, that some elements within Kunigalamma resist or survive hierarchical spatialization. Similarly, the context attaching to the contention that Kunigalamma is a form of Parvathi herself and/or that she is a demonic manifestation of Lakshmi confounds the impression that she is easily incorporated and legitimized within a Sanskritic framework. In fact, it can be argued that such an assimilation reveals how Kunigalamma becomes a ground for articulating the respective claims to dominance of the upper castes themselves.

Attaching her to a deity in the pantheon and distinguishing between her meat-eating and vegetarian habits, these operations serve to self-represent and rationalize the higher castes' worship of Kunigalamma to themselves. However, they cannot be taken to imply that she is thereby fully subordinated to a wider hierarchical system. The very fact that she is made the subject of so many ritual practices and rationalizations only serves to reinscribe her centrality within the community. In other words, far from being subjected to encompassment by a social whole, this autonomous goddess indicates a certain ambiguity in the putting together of the whole, even suggesting that the part (here, the local female deity) can have a definitive role in the making (or unmaking) of the space of the whole.

Extending this point in a gender direction, one could perhaps suggest that the inner, domestic realm that has been identified as a crucial feminine space in ritual practices is far from being a space determined passively and moulded from the 'outside'. It should be discernible that in asserting thus, we are, at one level, taking issue with the Dumontian scheme of purity–impurity, where women are placed lower than or in a position of inferiority in relation to men, primarily as a consequence of their proximity to 'polluting' bodily functions (for example, menstruation, childbirth). This context, however, is transformed when approached along a matrix invoking auspiciousness–inauspiciousness.[10] As Marglin (1985b) has indicated, the auspicious–inauspicious categorization is not only non-hierarchical, but also takes cognisance of the positive powers of women in the domestic and extra-domestic realms. The importance assigned to women derives from their ritual role as promoters and sustainers of a harmonious life and, indeed, as we have seen in this chapter, at different junctures of ritual activity, the woman is the prime mediator in invoking and sustaining relations between the human and divine realms. This is clearly brought out in the role of married women as mediators of purity and bearers of auspiciousness.

Making better sense of these rather contradictory positionings of the sphere of gender would require not just a multi-angular theoretical frame, but also a re-reading of the very space of the domestic/olage itself. I shall return to this theme in the last chapter while pursuing some implications for sociological practice in India.

Notes

1. But see the discussion of uru habba subsequently.
2. Fuller (1992: 90) has a point of clarification on this. According to him: 'Deities are...not differentiated according to whether they are offered puja, whose food offering is normally vegetarian, but according to whether or not they are offered

sacrifice. We must therefore focus first on the relation between puja and animal sacrifice, rather than the dietary distinction'.

3. Though there is a small Budabudike settlement (approximately fifteen households) adjacent to the Banyan tree, it plays no role during the jatre.

4. One of the major villages in the cluster, Hulaguppe is seen as the natal home of the goddess. She is said to have been first worshipped as mother-goddess there by some of the earliest settlers in the village, and over time became the *gramadevate*. The people faced a number of difficulties in building a shrine, however, and finally the task was accomplished at Maranahalli, a nearby village.

5. The episode was recounted thus: 'When Krishna was born and smuggled out of prison, Yashoda's baby daughter took his place. As usual, Kamsa ordered the death of the girl, but she was not destroyed and took another form—that of Adishakti or Parvati. When Krishna slayed the demoness sent to kill him, there was an uproar in the demon world. To pacify them, Krishna promised that ordinary mortals would worship them regularly, acknowledge their strength and evil powers, and plead that it be kept under check.'

6. Kunigalamma's procession leaves Maranahalli by noon, after worship and ritual crossing over fire. The horse-mounted goddess (it is believed that she journeyed from Kunigal on a horse) is placed on a pedestal which is framed by a wide, squarish wooden board. The idol is well-decorated, covered almost completely by flowers and garlands, leaving the head barely visible. The men carrying the goddess on their heads dance for several hours, with hardly a pause. Several currency notes are pinned to the board, which are supposedly muyyi payments made as vow fulfilments.

7. Cf. Fuller (1992: 200): 'The concern with finding a good husband partly reflects the fundamental importance of marriage for women in India. Marriage confers full maturity on females and transforms girls into *sumangalis*, personifications of auspiciousness.'

8. See Hegde and Niranjana (1994) for an exploration of this point and its possible implications for a discussion of the hierarchical principle. Several other suggestions are made in that paper, which have not been extended into this analysis.

9. This has often been described as the *sati* and shakti aspects, the former embodied by the benign spouse goddess and the latter by the autonomous single goddess. See, Gatwood (1985).

10. For literature on the pure–impure and auspicious– inauspicious axes, see Das, 1982; Khare, 1976; Srinivas, 1952. I shall traverse this ground more actively in Chapter Seven.

Six

In the Tracks of Women's Agency

Central to all discussions of the social field is the question of human agency. The latter, understood in relation to women's lives, has often taken the form of resistance offered against the exercise of power. Most such attempts tend to define women's agency in terms of a 'transformative capacity', meaningful only within a 'politics of change'. While this is definitely a crucial link that has been substantially articulated within feminism, it has also served to pre-define our engagement with women's agency. As a result, only certain kinds of political actions of women are seen as agential, revealing, at the same time, a deep unease about certain other kinds of women's acts that apparently blend into existing structures of patriarchal dominance. Interestingly, underwriting this perspective on agency has been the implicit delineation of space into public and private, with the former being described as an overtly political domain—as, in fact, the domain of power. The relative marginalization of women from this domain, consequently, gets interpreted as a sign of their power-lessness and lack of agency. To be sure, attempts have been made to read resistance into women's acts within the so-called private sphere itself (O'Hanlon, 1991; Raheja and Gold, 1996). But it is noteworthy that this focus issues from an alternative conceptualization of power—less as a unitary, monolithic structure and more as a sort of tenuous space, constantly fractured by the contestatory acts and gestures of subordinates. While women's agency would here inhabit a more cultural, even individual-centred, transactional matrix, the approach which defines women's agency as meaningful only within a 'politics of change' seems to locate it within a more pronounced structural ground. The two approaches, in effect, traverse divergent paths and offer, inevitably, partial accounts.

Rather than engage with their apparent incommensurability, my attempt here is to try and formulate an alternative ground from which to thematize women's agency. Such an approach consists of tracking agency primarily through its bodily locus, while also examining its intersections with sexuality. By doing so, I can also carry forward the engagements of the previous chapters. Spatial considerations, as I have argued, are basic, and, to the extent that my account will issue off the public, visible quality attached to disputes, I believe this will lend a further dimension to the charting of the logic of insides and outsides.

Disputes: Internal and External

Although the idea of domination–subordination is intrinsic to the concept of power, one could maintain that it is impossible to identify clear-cut bodies of dominators and dominated. Indeed, as a sustained engagement with the intersections of space and gender would reveal, the agents and arenas of power are variable and not fixed entities. As Foucault (1986: 98) has observed:

> ...(P)ower...is not that which makes the difference between those who exclusively possess and retain it, and those who do not have it and submit to it. Power must be analysed as something which circulates... (Individuals) are not only its inert or consenting target; they are always also the elements of its articulation.

Drawing on this insight, a further inflection into the edifice of the olage–horage demarcation could be attempted. Rather than posit the olage as a *dominated* realm—in relation to the horage as the *dominant* one—what is made possible is a dismantling of this very dichotomy. The tracks specific to women's agency in an everyday context must be approached in this light. But let me set up a passage through disputes.

Disputes within a community provide a very revealing picture of the fields of power, alignments and pressures structuring its space. My own interest in examining this ground derives from the constant invocation of socio-spatial coordinates in both the framing of disputes and their resolution. This is particularly true with matters relating to caste, property and the family, which are invariably marked by a reference to limits and boundaries. Especially during moments of conflict and strain, those spaces that are accepted and traversed as a matter of course tend to get high-lighted. Escalations of tension within the family, for instance, suddenly render visible the extraordinarily nuanced, yet clear-cut, spaces assigned to each member. There are strong moral reminders of what the place of the wife or the daughter-in-law should be, as well as specifications of the appropriate socio-spatial loci of their activities. While these may be read as points of power exercise, there is also, at times, a marked reversal: It may be the mother-in-law (or husband) who is accused of overstepping their bounds and, as the quarrel develops, the ground and limits of their behaviour get re-traced. More protracted clashes often culminate in further boundary lines being drawn, as in divisions of property or the places of residence themselves. In the latter case especially, the reference to limits takes a manifestly open form, evident, say, in the separation of hearths or in the erection of a wall dividing what was till then a commonly shared space. With caste-related disputes, the question of boundaries is more complex. It impinges on several aspects of the daily social interactions of individuals, such as the proscriptions on inter-marriage or on eating at one another's house. This assertion of limits is not just cultural but spatially enjoined too, a fact evident in the very physical structure of the villages.[1] Transgression of any of these spaces could snowball into a conflict between castes, or even a tussle between them for dominance. Needless to say, the more outright economic control and power wielded by the dominant castes over the lower ones cannot be ignored.

The point I wish to return to, however, is the processual delineation of spaces and its borders that is undertaken from diverse vantage points within the community. Nowhere is this more evident than in the ways through which disputes are framed and sought to be resolved. All disputes are inextricably linked to a notion of limits (*miti*)—about what one may or may not do, as well as what one ought or ought not to do—notions that form the hinge on which local morality often turns. Also, most disputes—be it over ritual matters, family conflicts over property, or quarrels between husband and wife—are described as 'internal' matters (*olagina vishaya*). Though these disputes invariably culminate before the village Panchayat, seeking external intervention (*horaginorna karesodu*) is lamented and seen as a blow to the integrity of the parties in contestation. In this drawing of lines between internal and external, and the consequent contouring of disputes, what is perhaps being sought is the consolidation of an already-fragile whole. But there is, it seems to me, a larger complex of meanings grounding this consolidation, and is particularly true of the various property-related and family-centred disputes that arise from time to time. Often, tensions between family members result in substantial disputations which may or may not reach the panchayat. Basamma, a poor Lingayat widow, commented:

> *Ayyo...do you need a mirror to see the boil on your own palm? In our house too there is trouble because of my daughter-in-law. Even my*

son has turned against me. That's why I've put up this wall partitioning the house. Why should I come between husband and wife?

Particularly among the middle and lower castes, conflicts with the daughter-in-law are often aired before the village Panchayat:

One should adjust and compromise. Otherwise of course there will be problems. Look at my daughter-in-law! She went to the Panchayat with stories that we beat and illtreat her...even had her own husband punished. She doesn't listen at all if we tell her that being younger than her in-laws she should keep quiet.

While these are but indications, the limits indicated in such disputes and the moral spaces they retrace are all about prescribed modes of conduct, about what is permissible and what is not. Disputes are thus one major axis along which a matrix of sexualization is invoked and sought to be reinforced. As implored earlier in Chapter Four, such a matrix enjoins a certain moral code upon women. It infiltrates their bodily and social practices so deeply, that acts and situations in the daily lives of women all respond to its normativity. Even the framing of disputes—especially family-centred ones—as 'internal matters' can itself be seen as part of the above matrix. Consequently, the act of a woman seeking justice at the hands of the Panchayat serves to confound the dominant delineation into inner and outer spaces. Nevertheless, it could still be maintained that the socio-spatial parameters enjoined upon women are somehow integral and come to be activated whenever women refuse (or fail to abide by) the spaces marked out for them.[2]

If these acts are to be thought of as manifestations of women's agency, it is also important to note the variability attached to it. Women's responses to the delineation of 'appropriate spaces' for them differ considerably across castes. For instance, in what seems a direct contrast to the disposition of the middle and lower castes, upper castes appear to exercise some restraint in taking family matters to the village Panchayat for resolution. Instances of bigamy do not always come up before the Panchayat, even if the first wife desires it, for it may be perceived as an interference in an internal private affair. Lakshmamma, a Vokkaliga woman, expressed a similar sentiment, saying that whatever the quarrels within the family, it should be kept private, lest the woman herself be blamed. According to her, 'wife-beating is really common, but nobody interferes. My husband may beat me all night, but by morning, we should be alright, and not broach the matter with others.' Complaining to a third person about the husband, it is insisted, would not only reveal the absence of a bond between them, but also lead to a break-up of the family. 'Take one problem to the panchayat, and you will have to take all matters there. So it's better to resolve all this within the house itself.' Both Subbamma and her daughter-in-law, belonging to a Brahmin family, voice their avoidance of 'public' spaces thus:

No matters concerning the family should go outside the threshold of the house. If we have a problem between us and speak about it to others, it will spread throughout the entire village. 'Did you know this about her?...the daughter-in-law is like this'...and so on. It will only anger family members if they hear this gossip. The best thing is to avoid all this at any cost. Nothing will happen if we remain indoors without letting matters cross the threshold.

Seen under this aspect, the fashioning of disputes is here witness to the emergence of a new dimension of women's space. Especially distinctive is the manner in which the olage—hitherto seen as women's space concerned with maintenance of the moral order—also provides the justifications for boundaries and controls internal to itself (in this instance, the household). Attention is thus drawn to how the parameters of power and dominance underlying the olage manifest themselves, as well as to their implications for a negotiation of the olage–horage continuum as a hybrid space of controls. What is paramount, nevertheless, is that these issues cannot be addressed monolithically but have to be mediated through a grasp of how social spaces are marked, gendered and reworked.

The socio-spatial grounding of disputes, then, provides us with one way of tracking women's agency in an everyday context. But, perhaps, the point could be made differently: that women's agency comes to be manifested precisely in the ways they insert themselves into existing socio-spatial grounds. The modalities of action and speech that characterize this insertion, while expressive of a matrix of inside and outside spaces, are also ways of negotiating within their contours (even, at times, destabilizing them).

Interregnum

The various pressures shaping and re-shaping the contours of the olage and the informal strategies deployed by women to deal with these may be glimpsed through the life story of Siddamma, a Vokkaliga woman in Doddaaluru. A rather reserved woman, she opened out over a prolonged acquaintance. The following account has been pieced together from conversations stretched across the fieldwork period.

Siddamma was the eldest child of her parents, with a younger sister and two younger brothers. When she was barely eight years old, her father deserted the family and was rumoured to have taken another wife. As a result, she began working along with her mother much earlier than she would have done otherwise. All she remembers of her childhood days is working from morning to sunset in and around Bannerghatta, which was where they lived then. 'We would begin early, carrying mud, cement or jelly stones. But as the day wore on, my only thought would be "when will the sun go down?"....'

She was married at the age of eleven or twelve. Since she did not come of age for about two years after this, she became an object of abuse and ridicule by her in-laws, who insulted her and called her an eunuch. She would live for a fortnight at her mother's house and for a month at her husband's, working for his family. They worked her too hard, she says, often denying her food and getting enraged if she looked into a mirror or even combed her hair. Soon after she came of age, her husband began to insist that she leave permanently for his house. He argued that there was nobody to care for him: 'Look at my stomach, there is nobody to cook for me.... Look at my back, there is nobody to scrub it...' and so on. However, after she moved, life at her in-laws house became increasingly difficult. Besides the back-breaking work and taunts she had to face, she was also not allowed to go anywhere even with her husband. Her eldest daughter was two years old when they decided to break with the family and move away.

Speaking about her relations with her husband, Siddamma claims that earlier they hardly quarrelled:

We were like two pigeons, content with what we had. People in fact used to say 'let us see you quarrel, for a change'. But now the whole thing has changed. After he started drinking, he has been frittering money away...the children could easily have had some new clothes. Now he has begun to deliberately find fault with whatever I do. 'There is too much salt in this' or 'is this the way to make the hittu?' ...it's all because of that Bhagyamma, I know.

She alleges that her husband is having an affair with Bhagyamma, a 'loose-principled' neighbour. As evidence of his roving eye, she recalls that he had offered to marry Siddamma's younger sister as well, even though she herself had borne him sons. He withdrew his offer only after relatives criticized him. Currently, their quarrels are either over the other woman, or because Siddamma complains that he is not bothering enough about the household and children. Her husband too uses the same argument to put her in place. He charges her with roaming outside the house and 'dressing up' to attract attention. He loses his temper if she talks with others, does not allow her to do wage labour, saying: 'You stay at home and do the inside work; I'll work outside.' According to Siddamma, this is a manifestation of his deep distrust of her, which also makes him strictly forbid her from going to her mother's place.

When faced with what she sees as her husband's excesses, Siddamma steels her resolve:

> *Have I ever done anything a wife should not? Do I talk needlessly to people, or go from house to house? I have not crossed the line drawn by him. Till now, I have bent according to his wishes, but its no longer possible for me to do that. I am not going to be soft and relenting.*

Her strategy in dealing with the situation seems to be to let others know about her husband's 'affair', so that gossip will control his behaviour to some extent. She prefers this to seeking justice at the Panchayat, for the latter would enable her husband to take her to task later for having made domestic concerns 'public'.

Siddamma's strategic choice of action, as we can see, is playing on the morality–space nexus—underlining her own 'morality' against another's infidelity and outwardly retaining the preoccupation with not making familial problems 'public', while at the same time striving to harness 'public opinion' (through gossip) to counter her husband's acts. In the process, not only does the inside assume a new (and wider) dimension, but one is also witness to how women's use of speech within the community could yield something crucial about their agency as well as their negotiation of spaces. It is this possibility that the following section addresses.

Gossip as Women's Speech

Attention to women's use of speech could shed light on the worlds in which they act and their intersections with prevalent normative structures. While theoretically one can refer to several aspects of women's speech—such as its diverse modes of expression, quality, contexts, content and so on—I would like to mention only some aspects here as preparatory to a focus on gossip as a neglected form of daily conversation among women.[3]

By and large, women's speech is not always direct, and their use of veiled accusations and criticisms, sarcasm, irony, and even silence in order to make a point in the course of a conversation or argument leads to their speech being typified (especially by men) in specific ways. It is dismissed as either non-serious and trivial, or feared as potentially dangerous, insidious and damaging. Both these perceptions acknowledge the 'difference' of women's speech, confirming the fact that it is seen as falling outside the dominant, 'more meaningful' communicative modes in the community. What I would like to underline here is that in the case of gossip its marginality rarely generates an oppositional discourse—a fact which makes an understanding of women's speech (and agency) that much more difficult. Clearly, one will have to look more closely and carefully at our notions of women's speech and agency, but I will return to this later.

The uniqueness of gossip, it seems to me, lies in the fact that in spite of being a marginal or peripheral discourse, it often works to define the social and moral boundaries of a community, ironically acknowledging and reproducing certain dominant societal perceptions. It is also distinct from other forms of speech for several other reasons. For instance, the 'stories' it weaves around persons usually take the form of rumours, which actually express doubts regarding what are paraded as 'facts' and seek to probe behind what goes on in everyday life. Through this very expression of doubt and probing, gossip situations tend to recreate and represent the bases of daily life and morality. It is also revealing that the sites of gossip often tend to be certain liminal spaces in the village— between people's houses, at the village pond, or the hand-pump. At times, it could even be at each other's homes when two or more women meet for some other purpose. But in spite of this quasi-underground nature, it succeeds in effectively policing the spaces inhabited by people. Another central feature of gossip tends to be talk about morality, especially deriving from contexts like the alleged sexual affairs of people, any transgressions of moral behavioural codes or deviations from accepted norms. Both in the issues it broaches and the ways in which it seeks to evaluate them, gossip creates an 'ambience of morality' amongst its indulgers, invoking and reinforcing certain moral perceptions of the community.

I will now present fragments from a 'conversation' between women in Hulaguppe in an effort to illustrate some of the ways in which women come to speak of the body and its location within a matrix of sexualization.

Sharadamma, Jaya and Seethamma are the main participants in this evolving speech situation. Sharadamma is musing over the problems of family life and attempting to console Jaya, whose husband has not returned home for over a week, ostensibly having lost face in the village due to some rumour involving him. Seethamma, who drops by, asks to be filled in on the details.

Sharadamma (Lingayat, 35–40 years, mother of five): It seems Ananthu (Jayamma's husband) had gone to Kalappa's house to borrow something. It was late after he returned from work in the city. When he was about to take leave of them, Kalappa's daughter and her husband suggested that he sleep there itself....

Seethamma (Achari, over 50, a grandmother): That is alright. But what kind of a woman is she to sleep in the same room as them? Couldn't she have slept inside, in the kitchen or inside room?

Sharadamma: Precisely. Had she been such a great pativrate, why did she have to go and sleep between the men? Look yesterday, two other men were here, along with him (her husband). Believe it or not, I didn't even go outside to relieve myself. I stayed in the inner room till they left. Oh, haven't we seen enough of her doings? But nothing has happened...that's why Ananthu is upset. Alright, let us say that while they slept, a hand or foot may have brushed her. But isn't that a situation she brought on herself? Why should she have suggested that he sleep at their place? And above all, sleep in their midst herself? What business does a woman have where men are gathered? If she is objecting to any kind of contact...can it be avoided today, tell me? We go on the bus, even share a seat with a man. Accidental contact can't be helped...it's up to us to conduct ourselves decently.

Sethamma: But why should Jaya's husband hide himself, as if something has happened?

Sharadamma: That's the whole thing. Ananthu is so upset because the charge is absolutely baseless. He did say that he has always considered her as a sister, and behaved accordingly. But he is so timid that he says he just cannot answer if somebody confronts him with this, as they are bound to. He is very soft...doesn't look a woman in the eye. But he should be firm about this. I am very clear about it. Ananthu is not at fault. That gives us a position from which to speak. If she has gone to the extent of accusing him of this, then we must reverse the whole thing. We must argue that it is basically her mistake. What woman in her senses would ask another man to sleep in the house at night? And was she a decent woman, to go out and sleep there? This is what we must ask.... I've been telling Jaya to go and bring her husband home. Let us not publicize the matter if he doesn't want it. We will call Kalappa here, to our house, and with my husband and some others, let us ask whether what his daughter did was right.

Seethamma: *Yes, let's place it before ten people and see what they say. Let's also call Jaya's mother-in-law.*

Sharadamma: *O, that woman will stand a mile away if there are men around. Like mother, like son. You know how young she was when her husband died. Yet she conducted herself in such a way that she didn't give others anything to talk about at all.... Even now, she draws her sari completely around her shoulders as soon as she sights a man.*

Seethamma: *These are difficult times, especially for a woman. One must be very careful.*

Sharadamma: *Even if you go around with a hand on your stomach, people will not stop commenting. And what will happen if you are not careful? That's why I don't like the behaviour of Kalappa's daughter. It's not right....*

Seethamma: *(Begins to bring up another rumour about the bachelor son of a rich landowner and an agricultural labourer.)*

The above fragment is both like and unlike a typical gossip situation on several counts. It is unlike one because, first, the participants are not passive spectators, uninvolved in the development of events. Rather, and this is the second point, it is their very involvement which prompts them to use the situation to chalk out a strategy for action. It is akin to a gossip situation in that there is a certain indeterminacy to it, where erring individuals are not clearly identified. This gives a strong impression that it is the perspective from which one views the situation that is decisive. It also suggests that who is actually at fault is but secondary, the primary issue at stake being the act of transgression of certain norms. It could be the man who is at fault; and equally, as the exchange between the women reveals, the fault could be reversed to place the blame on the woman. Such a reversal was being debated at that time, in the course of a seemingly innocuous and non-serious conversation between women. Perceived in this light, gossip, which is often dismissed as unimportant, can also be interpreted as an important ground on which certain strategies of action are worked out. In this case, the normal run of accusations would have resulted in the matter being brought up publicly before the Panchayat for resolution. But as the conversation shows, women's speech here serves to pre-empt this by preparing the ground for a 'private resolution', where nobody's honour is sullied. But the question really is whose interests are being served through such a resolution? (We could also note that Siddamma's use of gossip, mentioned in the section preceding this one, is somewhat more subversive, though she too aligns herself in relation to a dominant moral code in the process of setting herself up as the wronged.)

In arguing that gossip be examined as a ground for debating strategies of action, it must be clarified that I am by no means suggesting that such modes of talk and action are radical or subversive. My interest, rather, is to show how gossip participates in the sexualization of bodies and requires women to position themselves in relation to certain dominant codes. In the instance cited, gossip serves to highlight and reproduce certain socio-moral boundaries by emphasizing the spaces and behaviour appropriate to a 'good woman', even at the cost of blaming the woman herself for the lapses of the man. It is interesting to see how the female body is being spoken about and, in the process, sexualized. Strictures of morality are unfailingly seen as important and definitive of female identity, both by the woman herself and by others, though these strictures themselves may be differentially defined at any given point. For instance, when the married woman who figures as the 'object' of gossip here is chastised for not living up to the norms of femininity, the reference is to 'decent' conduct on the part of a pativrate, an inscription within marriage. She is not being charged with infidelity directly, but rather as having transgressed the bounds of pativratya. The ensuing talk would definitely have been slanted differently had she been an unmarried daughter. Either way, questions of morality are central to the production of a woman's space of identity, and gossip seems to underline this centrality. In this sense, gossip seems to speak from within a prevalent value system, reinforcing the rules governing thought and behaviour. Its potential subversiveness is thus often deflected due to a range of complex reasons, among which may figure women's perception of themselves as custodians of the group's values.

Another reason why such a gossip situation rarely offers resources for criticism is perhaps the other 'functions' it serves in a micro context like a village community. These have to do with its role as a circuit of information and as a site of a certain shared idiom—behavioural, cultural, whatever—which helps reiterate a sense of community, often in contradistinction to an outside(r). But what is noteworthy is that even in these other roles, an important preoccupation of gossip is with issues of female conduct and sexuality. To illustrate this point let me add some more angles to my earlier account. This draws on my experiences during the first couple of weeks of fieldwork in the village, a time when I was often the object of conversation (if not gossip) between people. A number of speculations were aired—some of which I came to know of much later—and others (in the form of queries, comments and veiled suggestions) I had to ongoingly engage with. A number of these concerned my work, of course, but more central were those hinging on my 'person': the fact that I was still single then (queries tinged with curiosity, genuine concern, or doubting disbelief, wild speculations about whom I was 'with', etc.), the problem compounded since, for several different reasons, none of my family ever showed up (were my parents not concerned about their daughter, an unmarried girl living away from home?), and further speculations triggered off due to my initial (rather naive) insistence that I would live alone in the village (for privacy and time, but these were reasons totally incomprehensible within the community, inviting comment, judgement and action). The issue was resolved when I began to see light and took up the offer of two families to send their young girls to 'chaperon' me. The other aspect which elicited disapproving comments was my 'appearance'—particularly my shortish hair, the absence of a nose-ring and often bangles too, sometimes even dress if I wore anything other than a sari. At such a conjuncture, being the object of gossip within the village community made me acutely aware and conscious of several codes of sexuality and morality in operation and, over time, these began to play a significant role in my own definition/self-presentation, in a way perhaps parallel to those of other women in the village.

I mention this only to underline the manner in which bodies are also shaped through speech which, especially in conjunction with other bodily practices, serves to confer an insistently feminine and sexualized identity. This is not to deny that at a more conscious level speech could also serve to register tensions within (or resistances to) these embodied identities; and it is in fact this very tension that I want to highlight as problematic when we refer to women's identity and agency. If we turn back to our description of a gossip situation, for instance, there is a constant harping on the trope of a pativrate—here appropriated not only in terms of fidelity, but also as one who is defined in terms of the 'inside', a sacrosanct space, free from violations—and alongside this iteration are hints of its apparent unfeasibility when women voice the dilemmas of negotiating this identity whilst traversing more public spaces (the 'outside'). The troubled relationships between women and the discourses/practices regulating sexuality, as well as the kind of surveillance afforded by speech forms like gossip also surface in a subdued manner in the reference to Ananthu's mother, who in 'disciplining' herself 'didn't give others anything to talk about at all', or in the observation that 'even if you go around with a hand on your stomach, people will not stop commenting'. What these overtures, resisting and submitting at the same time, signal is the extent to which women's identities, their senses of self and their acts are circumscribed within a range of bodily practices within the community.[4]

The Question of the Bodily Subject

Attempts to chart women's agency will, accordingly, have to be traced through the body, engaging with how women's bodies are both lived and imaged. That is to say, the concern should be with how bodies bear the marks of definitions of (female) sexuality and morality. The task is by no means an easy one, for what we are really engaging with is the question of how gender comes to be constructed, while simultaneously keeping in view the materiality of bodies.

The normative pronouncements embedded in discursive practices are not extraneous elements that come to impinge on a pre-given body, but are constitutive of that very body, marking it as female/feminine. It may be imperative, therefore, to enquire into what actually happens in the course of responding to these markers. Are processes so simple and clear-cut that we can speak of women as *either* upholding a hegemonic moral order *or* as deviating from it and resisting it (and, in that resistance, truly manifesting agency)? To arrive at such a conclusion would, clearly, be a distortion of the shifting grounds of, and engagements with, normative discourses. For women speak both from within the dominant discourse and from outside it, transgressing order at times and endorsing it at others. The quality of women's speech and action is therefore ambiguously located, not wholly outside or oppositional alone but also contributing to a replication of the moral order. In that sense, women's speech/agency is double-edged, manifesting a shifting trajectory with uneven effects and split along dual axes of reinforcement and transgression. These double inscriptions, while located centrally on the space of sexuality also suggest the various strands that go into the question of female subjecthood and its constitution.

Consequently, even as the impossibility of addressing agency without contending with its bodily basis is stressed, we would need, in addition, to clarify our understanding of the bodily subject. Subjecthood is certainly a complex question and much can be (and has been) said about it. But if we conceive of the process of subject formation as one of 'subjection', it would involve, at the very least, two aspects: one, as offering a locus of subjectivity, the space within which the singularity òf the individual engaging with societal processes is inscribed and, two, subjection as also involving submission, not necessarily forced but often, as Balibar (1994:9) puts it, 'willing obedience, coming from inside'. What this suggests is that the 'disciplining' of bodies within cultures is often unconsciously done where subjects play an important role in taking upon themselves certain self-definitions, and in giving expression to these in practice—whether through speech or act—may ultimately end up reiterating a certain moral tone. But while the apparent effect is one of the appropriation of a 'legitimate' discourse and its reinforcement at the structural level, a lot could be happening at the daily level, where subjects, as singular persons, could well be engaging in transgressive acts as well.

It may be useful to remind ourselves of the root meanings of the word 'act' itself (as verb) and ask what it can imply for our understanding of subjection and subjectivity itself. The *Oxford English Dictionary* has this to say on the word: 'to carry out or represent in mimic action;...to perform...to simulate, counterfeit...a performance...pretence (of being what one is not)'. Accordingly, if one were to inflect this meaning into the contexts of female agency discussed, it could be suggested that women are, through miming and appropriation of a moral/normative standpoint, reinforcing an existing value framework. But, on the other hand, their speech-acts also seem to gesture occassionally towards a transgression of those values, particularly through the inversion of constitutive definitions. In the earlier gossip situation, while there are women condemning another for an alleged transgression of 'morality' as they perceive it, there are also suggestions of how difficult it is to live as a woman, indexing thereby the burden of negotiating both the imposed and taken-upon self-definitions. What this would implore is a view of female agency as, at once, rooted in bodily identity and also engaging with a socio-spatial order that constitutes and defines it. This engagement can be deeply contradictory—reinscribing a dominant order, on the one hand and, on the other, querying its very basis.

Of course, this transgression–reiteration axis need not exhaust the quality of women's speech-acts. Rather, it encapsulates the boundaries within which female sexuality, female speech and action become comprehensible. Within the gossip situation, what is spoken about, how it is being spoken, and what strategies are being devised are all illustrations of a tension between the disciplining of the body and its transgression, where the over-running of desire within a spatial framework is seen as eliciting a regulatory discourse as well. Gossip is thus both more and less than the socio-spatial order—more because of its effect of reiterating it and less because it is also about transgressions of that order.

Grappling with questions of women's agency must, then, involve negotiations on the terrain of the *bodily* subject. In particular, one will have to contend with how women *live* their bodies, live *in* their bodies and *with* how they contend with *what cultures make* of such bodies. The experience of such a space is contained within what I have disclosed as a matrix of sexualization which, while issuing from the body, condenses into codes of community and necessity. As each of the chapters have ventured to state, such a matrix is premised on a distinctive spatialization of social life. The fluid demarcations of inner and outer spaces come to ground (and are indeed grounded by) perceptions of gender that also mark and differentiate bodies as properly or improperly feminized in relation to existing normative codes. While the coordinates of such a matrix of sexualization/spatialization could be derived from various institutional discourses on the body—legislative, governmental and so on—I have tried to formulate the basis for an ethnographic retrieval through socio-spatial parameters. The olage–horage affords a unique point of entry into the matrix of sexualization/spatialization as retraceable within a village cluster. It also lends a significant spatial dimension to the concept of gender itself, suggesting a possible vantage point from which to invest gender itself with a determinacy often absent in constructivist accounts. Theorizing these domains could have implications for conceptualizing both gender and sociology in India, and it is towards beginning this task that the remaining chapters of the book turn.

Notes

1. The appendix has the details.
2. Though I have confined my reference to disputes involving women, the same would be true of any dispute (say, those of caste, ritual or any other matter).
3. Speech as a form of self-expression is often unproblematically related to women's agency, just as the absence of speech is associated with powerlessness (see Sundar Rajan, 1993). Note, that in examining gossip as a form of women's speech my attempt is to complicate the terms of this equation.
4. Recall fragments of the discussion in Chapter Four. The accretion of meanings exhibited by references to the stomach—or binding the body in different contexts—signifies also an imperative to control and spatialize.

Seven

Concomitants and Implications

Tracing the conjunction of space and gender in the practices and discourses of femininity and sexualization has been central to this book. Aspects of the spatial narrative governing forms of life within a village community have been specified. My task, in this chapter, is to recall, or point out, dimensions of this narrative in a bid to overcome the reluctance to theorize the significance of the spatial register.[1] This will, by necessity, consist of wading through certain cognitive frames embodying Indian sociology and anthropology. These largely dichotomous frames—such as the opposition of pure and impure, auspicious and inauspicious, right and left—have overseen the institutionalization of the discipline. Given the strong hold that these frameworks have had over the study of Indian society, it is imperative to ask whether they can convey the fluidity so characteristic of the social field. Indeed, what difference would a processually grounded, spatial account of Indian society make? The contemplation of this context looms large in the ground that I shall now traverse.

Founding Polarity

Looking back, the debate about hierarchy and its place in Indian society, foregrounded above all by Dumont, has been among the axial questions fashioning the specificity of Indian sociology. Though I will not directly grapple with this issue here (but see Bailey, 1957; Beteille, 1992 for critical views), Dumont's significance is nevertheless undeniable given the large number of sociologists and anthropologists who have been engaging (either in contention or in agreement) with his isolation of purity and impurity as the founding polarity for understanding hierarchy in caste/Hindu society. Using Dumont as a point of reference, one could ask: What does this polarity signify? More centrally, we need to address the ideas, values and acts of behaviour codified in and by this polarity, even as the efforts to displace it are consolidated. My concern, to be sure, is not so much an enumeration of detail, but the offer of a recognizable portrait, and thereupon to direct attention to other insufficiently articulated principles and domains.

The core of Dumont's work consists of the attempt to 'isolate the pre-dominant ideological note of a social system' (Dumont, 1980: xxi), to 'reconstruct' it so as to uncover the fundamental principle underlying Hindu society. To this end, he focuses on caste as a system of ideas and values, with the conviction that 'castes teach us a fundamental social principle, hierarchy' (Ibid.: 2). Governing the working of this system, according to Dumont, is a structural principle guided by the opposition between the pure and impure: 'The opposition of pure and impure appears to us the very principle of hierarchy' (Ibid.: 59). This principle, seen as suffusing caste society, posits the ritual or sacred realm as primary, separate from a 'profane' politico-economic sphere dominated by issues of power, territoriality and so forth. A concept of encompassment is held to constitute the very essence of this hierarchical relation, encoding as it does the relations obtaining between the two realms and in the context of which the politico-economic realm is encompassed by, yet set apart from, the ritual realm. Or, in other words, where questions of power are surrounded by, and/or subordinated to, the ideology of the pure and impure.

This postulation of the ritual realm—or purity-impurity as the central axis animating caste/Hindu society—is embedded in a number of pre-suppositions, whose implications have been critically engaged with by scholars over the decades. Though these critical voices are far from homo-geneous, the most insistent has to do with the Dumontian subordination of power and its relegation to the politico-economic realm. This criticism has meant contesting the Dumontian claim of the pervasiveness of the purity–impurity polarity. While some point out that these values are mani-fest only at the extremes of the caste hierarchy (Srinivas, 1978), others seek to highlight the complex folds attaching to the pure and impure by focusing on the inter-penetration of the ritual and politico-economic realms and underlining the role of the dominant caste (or, as in the past, the institution of kingship). The works of Raheja (1988) and Dirks (1987) are perhaps emblematic. Quite ironically, it appears as if the very force of Dumont's argument has raised questions regarding the place of other domains, principles and processes in Indian society, triggering off studies imploring—if not always in endorsement—an alternative repertory of categories and frames.

There are a plethora of such frames used to explicate and understand Indian society today, most of which criss-cross with one another. Though their theoretical and institutional affiliations are diverse, they have broadly sought to affirm the limits of binary frames of reference and the need to theorize domains which are not directly subsumable under the purity–impurity polarity. One may attempt to orchestrate this medley of emergent voices thus.

The Limits of Binarity

The category of 'auspiciousness', among others, has been analyzed as a crucial principle ordering the social world and the flow of events in it. Though this principle is discussed primarily in oppositional terms, that is to say auspiciousness as a counterpoint to the emphasis on purity, it is necessary to ask whether these categories are to be approached as mutually exclusive. Perspectives on this issue vary. On the one hand we have Srinivas' early study (Srinivas, 1952) highlighting auspiciousness as a major cultural category in addition to purity–impurity. (It is also a fact that Srinivas himself later collapses distinctions between the two, creating over-arching categories like 'good–sacred' and 'bad–sacred'.) On the other hand, Dumont and Pocock respond to such moves by seeking to re-assimilate auspiciousness into purity; the latter, for them, being the fundamental principle. There also obtain other viewpoints which, building on the contention that 'auspiciousness has received far less attention than...purity' (Marglin, 1985a: 1), postulate crucial differences between the two principles. Likewise, attempts have been made to shift scholarly attention to other vital domains and aspects of Indian society. Madan (1987: 12), for instance, has observed that '...the well-established emphasis on the study of purity–impurity, though undoubtedly of fundamental importance in the study of Hindu society, does not provide by itself a sufficiently adequate framework of values for the study of domestic life', adding that the latter domain is characterized by 'auspiciousness, purity and moral equipoise' (Ibid.: 1).

Clarifying the notion of auspiciousness has thus been one of the routes to displacing an emphasis on purity and impurity. Another alternative is suggested by Das (1982), who seeks to develop a conceptual apparatus oriented towards understanding aspects of Hindu caste, ritual and myth. She explores systems of classification based on spatial categories such as right and left, as well as in terms of the cardinal points (north, south, east and west), while arriving at an analytical axis aspiring to go beyond the pure–impure opposition. In refusing to subsume the right–left opposition under purity–impurity, she is led, however, to setting up a new binary (auspiciousness/life/right–inauspiciousness/death/left) in ways that could be, in their turn, critiqued.[2]

To be sure, most of these clarificatory attempts are laced with a sharp awareness of the insufficiency of translating the richness of indigenous terms into hold-all concepts and of the distortions that this entails. Perhaps in response, the trend is now towards using indigenous concepts themselves while documenting the variety of arenas in which these categories come into play. Though this should ideally provide fertile ground for reconceptualization, the emphases laid and interpretive glosses provided are so diverse that there still emerges no coherent framework for understanding everyday dynamics. One could illustrate this diversity by referring to the concepts of *subha* and *suddha*. Subha (or 'auspiciousness'), as detailed by Madan, has reference to 'time and temporal events in relation to particular categories of people' (Madan, 1985: 12) and is 'also associated with places, objects and persons connected with the (above) kinds of events or actions' (Ibid.: 13). Subha thus is that 'absolute value which manifests as a quality of events in the lives of human actors (*patra*) and involves the dimensions of time (*kala*) and space (*sthana*)' (Ibid.: 17). In contrast, purity or suddha (and *asuddha*) is seen as having no reference to events, being 'attributes of animate beings, inanimate objects and places...' (Ibid.). In this account, what is immediately apparent is that because the invocation of auspiciousness as value is linked (though in the abstract) to co-ordinates of time and space, it retains the possibility of situational translation. Alternatively, since purity–impurity are defined as intrinsic qualities, they forfeit this possibility.

The non-coincidence of the values of auspiciousness and purity is pressed home in other studies as well. Here, purity is viewed as pertaining to the ordered hierarchical domain of caste, whereas auspiciousness is distinctly non-hierarchical and free-flowing, signifying a state of well-being and happiness, fertility and growth, progeny and prosperity (Marglin, 1985b). And though the coupling of the terms 'auspicious–inauspicious' suggests a binary opposition, there is the suggestion that no exclusivity exists in practice; indeed, that there is a co-existence or intermingling of both. As Marglin's work indicates, persons such as devadasis, events like birth and even rituals like ancestor worship indicate that 'the opposition between auspiciousness and inauspiciousness is not an exclusive binary one, but one that lacks a fixed boundary between poles' (Marglin, 1985a: 79).

If the fluid complexity of events and categories is what is being high-lighted, then what is offered in place of a binarist model? Since the answer or solution to insistent dualizing cannot lie in monism, some suggest a three-dimensional graphing where variables are anything but static (Marriott, 1990). This is a view, characterized as 'ethnosociological', which argues that the social world is not only layered but may be viewed from multiple angles; that constantly altering positions and perspectives in the transactionality of daily life dispels any fixity, even approximating to what Malamoud (1981) has termed a 'revolving hierarchy'. A number of recent studies endorse this flexible model of Hindu society, one such being Marglin's (1985b) attempt to represent the royal and wifely do-mains in its changing relation to status hierarchy. Similarly, Raheja's (1988) description of the shifting nature of caste relations—through a focus on pre-station patterns and the ritual role of the dominant caste—offers an alternative view of caste/Hindu society to the one based on a single principle of purity–pollution alone.

Inflecting Into Space

It should be discernible that these studies, despite seeking to dislodge the purity–impurity polarity, neither issue from nor strictly lend them-selves to a characterization of space and spatiality. This lack, to say the least, is surprising, since space/spatialization could have provided an alternative ground for anchoring the very flexibility being sought. Perhaps the larger structure of assumptions in terms of which space is approach-ed—and which, as we saw in Chapter Two, has to do with the highly questionable association of space with stasis—is responsible.[3] Besides, it can be mentioned that a comprehensive account of the situations attach-ing to women's lives remains peripheral to these disputations, a point which, again following the considerations brought forward in Chapter Two, can be linked to the absence of a critical perspective impinging on space/spatialization. I have in a series of chapters sought to bolster and carry forward this claim, illustrating by means of an ethnographic context what such a perspective could entail. Let me here continue to dwell on a more cognitive/theoretical matrix issuing off Madan and Marglin and implicating their accounts of domesticity and auspiciousness respectively. I shall in the final section return to the olage–horage demarcation ground-ed in and by my ethnography.

In any attempt to relate space with gender, what would be immediately apparent is the centrality of women to the formation and continuation of the domestic space. While this identification of woman with the domestic could be explicated in terms of her material contribution to the running of the household or her tasks of childcare, a more prominent axis would be her association with the quality of 'auspiciousness' as well. This is, however, a relation that has till very recently received no systematic theoretical elaboration in Indian sociology and anthropology. The absence is palpable, for instance, when we read the otherwise detailed essays by Madan bearing on the categories of subha and suddha in Hindu culture (Madan, 1985) and on domesticity and detachment (Madan, 1987). The latter essay especially intends, through delineating elements of the ideology of the householder, to speak of the place of the *grhasth*s in the daily life of the Kashmiri Pandits. Accordingly, the central concerns of domestic life are seen as the proper execution of one's social roles/dharma, including the following of customs, performance of daily rituals, routines and samskaras. What is striking—and is even recognized as such by the author—is the lack of reference to women within this ideology, an absence attributed by Madan's informants to the ritual impurity of women, their dependence upon men in several matters and the confinement of women's activities to the kitchen. Madan interprets these attributions as underlining the necessarily hierarchical and encompassing relation of men vis-à-vis women. More importantly, even as the inherent reversibility of hierarchical relations is acknowledged, this does not translate into the possibility of developing an axis of theorization which would frontalize the sphere of women's lives and women's spaces. Indeed, a closer and more concerted focus on the practices of the daily world would demonstrate that women, although sidelined by the purity–impurity axis, are undeniably central to the household/domestic realm on other counts.

The theoretical implications that this could have for a study of women is noted by Marglin (1985b), particularly in her postulation of the principle of auspiciousness–inauspiciousness as non-hierarchical. The non-feasibility of a singular perspective on social dynamics draws her to propose a flexible model, whereby the dominance of the wifely (and kingly) domain can be rendered equivalent to the man (and Brahmin) and even reversed depending on the variability of the viewpoint being adopted—whether that of impurity or auspiciousness. That is to say, while women are 'inferior' due to their association with impurity in a hierarchical scheme, they are to be approached as the sources of prosperity and well-being along a more non-hierarchical axis, and hence as essential to the household or kin-group. (I must hasten to mention that even as Madan acknowledges women as bearers of auspiciousness, this finds no place in his effort to explicate domesticity via the ideology of the house-holder.)

The possibility of women becoming visible subjects through the domain of auspiciousness need not, of course, entail that the significance of the spatial has been grasped. In fact, in this regard, one might note that while Madan dwells upon the imperatives and intricacies of relating auspiciousness to the flow of time, its relation to space remains largely undeveloped. Marglin too discusses notions of auspicious–inauspicious as inextricably linked to ideas about time. The reference to almanacs and the suggestion that women's auspiciousness and fertility are associated with natural cycles of renewal bear out this emphasis. And yet, although the spatial is never theorized as such, her own discussion of devadasis and of temple rituals seems to provide some indications of its significance. Thus, for instance, devadasis, in spite of being considered 'auspicious beings', are not allowed within the inner sanctum of the temple. This avowedly spatial operation bristles with questions of how sexuality and impurity, auspiciousness and chastity are differentially articulated within a given space. Needless to say, the implications of such a spatialization cannot be contained within temple precincts alone and extends beyond into the everyday lives of women. In the context of the latter, as I have tried to disclose, space—both physical and social—is perceived, organized and arranged as a series of insides and outsides. Of course, the transformations that this spacing could effect for the principles which structure distinct ways of life remain to be traced. Neverthe-less, rather than speak of ways of life within communities in terms of qualities (such as auspiciousness and inauspiciousness) or principles (such as purity–impurity), what is being implored here is an attentiveness to the nature of their shifting grounds. In a word, inflecting into space, but also inevitably ranging across gender.

The Olage–Horage Demarcation

The matrix of olage and horage, as my ethnographic rendering has sought to demonstrate, pervades and anchors the conduct of life and social practices in Aaladapalya. The demarcation and criss-crossing of inner and outer realms function not only to cut up and traverse grounds both physical and social, they also oversee the definition of gendered domains and identities. It has been shown how the olage comes to be set up as a moral, sacrosanct space in relation to a series of horages and how women's acts and lives are assessed in relation to these spaces. This spacing obtains within a variety of spheres, be it in terms of a screening of the relative work opportunities of men and women, across a gamut of ritual activities, or in the manner in which disputes are spoken about, framed and resolved.

One could also clarify at this juncture that the olage–horage taxonomy as a way of mapping spaces does not echo the domestic–public model which has often been used to account for the productive, reproductive and social roles of the sexes. A central assumption of the latter has been that women by virtue of belonging to a domestic domain centering around childbearing, rearing, cooking and other maintenance work[4] are excluded from the public domain, usually seen as involving political and economic activities (see Sanday, 1974). Since the public domain is associated with the exercise of power and control over persons and things, the domestic, in lacking these qualities, becomes by definition a subordinated space. In re-casting space (and its negotiation) as multiple and relational, the olage–horage taxonomy renders such oppositions irrelevant. There are continuous shifts and transformations in how these spaces are characterized, since they acquire and shed meaning according to context. A significant point then, which cannot be over-emphasized, is that the olage-horage demarcation is never a fixed one, nor does it stabilize across contexts, in spite of the tidiness that the words themselves may suggest. Rather, the boundaries of these spaces are constantly being specified and re-specified, contextually and in practice. In speaking of the olage as women's space, therefore, one must be careful not to miss the fluidity, the ever-changing contours of that space: it could signify the safety of one's own village when pitted against another village or town; it could refer, in the context of inter-caste relations, to the shared homogeneous space of one's own caste or kingroup; or it could even connote the intimacy of one's own hearth vis-à-vis other households. Depending on the dimension of the social being talked about, olage could signify either the immediate family, the extended kin/caste group, or the village (uru) itself—the determining factor in each instance being the horage it is contrasted with.

Attached to the shifting locations of the olage and horage are certain unambiguous qualities, though even these often come to be contested or tempered with more situational imperatives. Thus, while the olage (whatever its point of reference) is generally perceived as a moral realm, providing security and inviolability to the persons occupying that space, we have also noted how it often becomes a site inscribing dimensions of power as well. Further, while olage and horage are used overwhelmingly to describe women's spaces, acts, qualities and the like—and can therefore be of use in comprehending how femininity/sexuality is lived and spoken about—the functionality attaching to this categorization of spaces bears on other realities as well. There is considerable evidence of how entire caste groups or villages use this spatial terminology to demarcate physical boundaries, articulate ritual differences and construct social identities as well.

In drawing attention to this matrix which underlies the everyday activities and practices of people, however, we need to recognize that behaviour is not always or entirely governed by rules that are consciously articulated. In other words, there could be ways of living that are mechanically and almost unthinkingly followed, often being coded in the various practical taxonomies of the body. They are rarely the product of rational thought, yet locate and embed individuals firmly in the world by assuming a certain normativity and force that needs explanation. Bourdieu (1977), among others, has sought to understand the bases of such modes of behavioural practice, particularly the regularity and, at times, improvisatory quality that mark human acts. He characterizes this ground as the habitus, a system of mental and bodily dispositions inscribed into the constitution of the self. What our own analysis has sought to suggest is that the habitus is a precipitant of a matrix of sexualization, and that the spatial delineation of olage and horage is at once the medium and outcome of this matrix informing community. Traversing the diverse guises under which olage–horage is invoked is thus crucial not only in the understanding of lay practices, but also because it provides a readily accessible idiom for engaging with everyday life.

Recognizing the Matrix

Thinking back, it appears that the experiences of fieldwork alerted me in many ways to the significance of space in social interaction. What was then a nagging sense of doubt and unease about the shift in social codes and spatial registers that was necessitated the first time (and each time) I entered the field, now seems more comprehensible. At the beginning, however, Aaladapalya was a place external to me which I sought to 'enter'—a movement from outside inward—which necessarily raised a number of questions: those of boundaries and the clear-cut lines of separation they entailed, of how one crossed boundaries, and most centrally of what such a crossing and re-crossing meant. As I grappled with these questions and with people's perceptions and ways of living, space emerged as a substantial factor grounding social life in ways that my training in sociology had not prepared me for.

Aaladapalya, comprising as it does of a cluster of villages and hamlets, was itself a heterogeneously ordered realm. Doing fieldwork in a cluster was definitely different from studying a single village, since the reality of boundaries was frequently encountered while negotiating the cluster. During this time, I chose to stay at Budkepalya for two reasons: one, because my first contact in Aaladapalya was based there; and two, for more practical conveniences (the community hall there offered some basic amenities in the form of tap water and an enclosed bathroom). I stayed in a room in this building, which functioned as the local *balwadi* during the day. Without my being very conscious of it at that time, this very choice of place located me within the community and cluster in specific ways. In the context of the group of villages comprising Aaladapalya, the hamlet I lived in (Budkepalya) was a peripheral entity. It was a relatively new settlement, dominated by Budabudike families, and known for being the site of interventions by various social work agencies. If at all one could speak of a core of the cluster, this was not it. In Budkepalya itself, the fact that I was staying independently and not with any family conferred on me a certain liminal status—of one neither fully inside nor entirely outside, but moving in and out, hovering on the margins so to say. While at the other villages I was designated as the one from Budkepalya (*palyadoru*), at the hamlet itself I was 'the one from Bangalore' (*bengalurinoru*). At least at the outset, nowhere was I fully part of an 'inside', and having to steer through these shifting identities sensitized me somewhat to delineations of olage and horage.

As an urban-born and urban-based person, there were several aspects of my bearing and perhaps even thinking that set me apart as an outsider. At times, even appearance marked me as different. But most revealing of all was people's perception and evaluation of the fact that I was, then, 'still unmarried'. 'Why? Were my parents not concerned? How did they allow me to come so far?'. That I was unmarried also served to occasionally pre-define the parameters of our interaction and conversation, with certain issues and questions being considered out of bounds. Underlying their responses to me, especially concerns about my having to move around during the course of my work, was the implicit suggestion that this was something 'not done'. It also brought home to me the parameters circumscribing women's lives, especially the ideas and values that contributed to the sexualization of bodies. It was through my own bodily location and traversals of the field that I became aware of the habitual bodily practices of other women in the field—their modes of speech, dress, spaces of work, as well as their negotiation of these arenas. This is what I have designated as a matrix of sexualization. In other words, a consciousness about such a matrix comes only when one treads the margins between inner and outer. As an insider, nothing is verbalized, everything is habitual; as an outsider, everything is strange, alien, but the one who mediates between the two worlds recognizes the matrices for what they are: *in space, across gender.*

A Further Formulation

The complications offered by the socio-spatial matrix of olage and horage give rise to challenging questions which are not altogether easy to answer. A fundamental objection could be that the taxonomy has been extracted from daily processes and that it hardly corresponds to the assertion of a concept; even, that while an indigenous register directed at mapping and grasping social life in Aaladapalya has been offered, the task of effecting a theoretical summary of such an avowedly spatial register still remains. Let me here venture a formulation.

Much of the specificity of Indian sociology could be recovered in the debates surrounding hierarchy and its place in Indian (caste) society. While this by no means reduces all realities to caste, or delimits the range of concerns and emergent interests in sociology today, it is a fact that caste has been considered a major index of cultural specificity/difference. As I have sought to disclose, the two important analytical frames that address these domains of caste and ritual, namely, the purity–impurity principle and categorization in terms of auspicious–inauspicious, seldom take space into account as a significant variable in social analysis. Again, despite there being notable studies demonstrating the centrality of women to the delineation of caste boundaries, an elaboration of these insights in a manner that could yield new conceptual resources for framing Indian society has not been forthcoming. In other words, the possibilities of gendered analyses have been highly delimited, provoking a search for other ways of framing the questions.

A nuanced deployment of the olage–horage matrix, I am suggesting, makes it possible to effect an important shift in our modes of analyzing both women's lives and the societal structures that surround them. The olage–horage delineation draws attention to a grounding matrix (both social and spatial) that is visible in one's bodily practices. On the one hand, it exemplifies the manner in which the woman or the female body becomes central to the drawing of boundaries within and between communities; on the other, it also demonstrates how certain notions of space and movement are written into definitions of the female body, informing the ways in which women's lives, acts and speech are structured. The matrix itself is no monolith impinging uniformly on all women. Considerations of caste, economic background or even religious affiliations insert themselves into, and consequently refigure, its parameters, making for an extremely complex mapping. The rules of olage–horage, consequently inflect differently under different conditions (say, in relation to upper-caste and lower-caste women). Yet, in spite of a number of such variations in practice, there is a broad commonality in the *principle* that informs their bodily practices. A mapping of these variations and differences is an urgent and challenging task, but one that can only follow from a prior recognition of the centrality of such a socio-spatial matrix to the sexualization of bodies. It is this latter point—rather than the former task—that I have tried to enunciate in this book.

As we have tried to demonstrate, the socio-spatial matrix forms the hinge holding together a rather fluid and yet determinate terrain. To be sure, there is nothing indefinite or fuzzy about the demarcation of inner and outer spaces, but what is striking is that their points of reference are constantly shifting. Perhaps better described as *sets of relationships* between entities, what this matrix facilitates is a mode of conceptualizing how the social world is negotiated (as opposed to, say, being strictly slotted or compartmentalized). Besides, its conceptual grounding is distinctly different from forms structured by a certain binarism. Neither an intrinsic essence nor trait (like purity–impurity), nor a quality that can be attached to people and events (like auspicious–inauspicious), the olage–horage is suggestive of a field regulating relations between entities that have a socio-spatial existence—be it gendered persons, families, castes, or even entire villages/localities. To reiterate: unlike purity–impurity and auspicious–inauspicious, the spatial axis foregrounded by the categories olage and horage is not marked by any distinct ideological connotation. The reference is less to distinctive ideas and values than about how, in the processual delineation of spaces, bodies and bodily moralities come to be constituted. Furthermore, as my detailing of lived practices in Aaladapalya has suggested, spatial considerations are an important medium through which larger values and ideologies impinging

on the body (whether these be chastity, morality, pollution or auspiciousness) are realized and engaged with on the ground.

It must be noted at this point that even the binary axes of purity–
impurity and auspicious–inauspicious do bear a reference (albeit discrete)
to the body; indeed, they presume the existence of bodies. And yet, it
seems to me that their concern is less with the body per se than with the
essential qualities or values that are attached to bodies. Alternatively,
the spatiality afforded by the concepts of olage and horage indicates that
these qualities/values are not eternal givens but assume the meanings
they do only when bodies/persons inscribe themselves in space in particular ways. At once material and cultural, substantive and discursive,
the matrix offers a more adequate mode of conceptualizing agency and
community within Indian society. It also exemplifies the point that space
is not a mere site of action; in fact, the social is always constituted through
the spatial, with the experience of lived bodies forming the basis of a
mediation across these spheres.

But there is a further complication that needs to be addressed. It is not
as though the socio-spatial matrix only affords us a glimpse of the dynamics of everyday life; it also indexes the manner in which societal reproduction takes place. The retracing of a normative grid produces gendered
bodies no doubt, but over time, this production is also a reproduction of
ways of living and seeing. Thus, while the concern that a foregrounding
of spatiality could deflect attention away from a temporal axis is by itself
a genuine one, it appears to be misplaced in the contexts being mapped,
because the manner in which the socio-spatial grid is being frontalized
and captured here defies any association with stasis; indeed, the fluid
contexts through which the grid materializes, as it were, seem to express
a dynamic core that could well constitute a condensation of the temporal.
In other words, a fundamental recognition that is built into the olage–
horage terminology is that as contexts and circumstances change, the
modes of negotiating space too are transformed. Unlike a conventional
space–time dichotomy that allows itself to be easily mapped onto a static–
dynamic opposition, a persistent focus in this book has been to develop
a perspective—drawn from an ethnographic context—that views space
as an active–constitutive grid. This is neither to place the field/community
outside a temporal framework nor to deny the significance of structural
transformations at various levels, but rather to register these changes
through a negotiation of space by gendered bodies.

As we reflect on this mutual shaping and re-shaping of bodies and
spaces, a further question presents itself—of how to draw on this everyday

vocabulary (of olage–horage) and re-cast it onto a *conceptual* plane into terms that will explicate some of the principles on which women's lives in society rest. Throughout the book I have tried to demonstrate how the olage–horage vocabulary captures the myriad ways in which femininity is lived and experienced; through it, I have also shown how physical spaces as well as a range of structural contexts (for instance, the economic or the religious) are gendered. As I see it, development of the conceptual resonances of the olage–horage matrix cannot be divorced from the task of elaborating a suitable concept of gender. It can, in the process, be noted that a substantial part of the social construction of gender (especially the shoring up of notions of femininity) hinges on how female bodies negotiate everyday spaces. Women's bodily experiences (for example, of menstruation), related cultural practices (rituals or pollution rules) as well as economic activities are all mediated through distinctly spatial concepts. Again, as suggested earlier, olage and horage go beyond mere physical demarcations of space; in carrying a load of social and symbolic meanings, these also translate into a matrix of sexualization within which ideas of femininity and female morality come to be articulated and assumed. While acknowledging, in broad terms, that gender refers to the social organization of differences between the sexes (Scott, 1986; Connell, 1987), what is being emphasized here is that space—in its various guises—emerges as a vital idiom or medium through which these (and other) cultural differences are articulated and signalled. Olage–horage thus not only demonstrates how gendered bodies are circumscribed, it also focuses on a pervasive logic underwriting wider social processes as well.

Whether it be the sexualization of female bodies through bodily and cultural practices, or the gender asymmetries in accessing resources and power, the matrix of spatialization opens up a major route for gender analysis. Using olage–horage as a heuristic tool also enables us to transcend the culturalist versus materialist impasse in gender studies. The resultant intermeshing of cultural and material strands is effected, in my analysis, through the gendered body. Moving in and across space, such a body materializes the spatial principle while also lending a substantive dimension to the concept of gender itself. It allows us, in other words, to argue that though gender is a constructed reality, the body is central to its delineation. As I will show in the chapter to follow, there has been, until very recently, an under-theorization of the body in social as well as gender analyses. What the demarcation into olage and horage facilitates is a focus on that space of extension issuing off and folding into the body in a way more direct and concrete than has been possible thus far. I shall now, accordingly, turn towards a consideration of the prospect of an emergent sociology of the body.

Notes

1. It may yet be claimed that in this bid to overcome a reluctance to theorize a spatial register, the dimension of time or temporality has been dispensed with. Thus, Ram (1999) in a recent review has this to observe on some of the schemes privileged herein: that an exclusive focus on a spatial grid makes for a synchronic analysis where attention to contextual differences is confined to spatial variations, obscuring the manner in which contexts are shaped and negotiated over time. Perhaps. But a more closely orchestrated response would require that we come to terms with the scheme of this classification, that is, the juxtaposition of 'space' and 'time'. In the absence of such an engagement, this criticism could merely be reiterating the conventional association of space with statis, status quo or even a lack of politics, and temporality with the obverse of this. One could begin to argue, instead, that time and space are not external structuring principles but the very form and texture in terms of which life experiences are organized; they are indissociably related and not oppositional. To re-assert a spatial perspective is not to banish the temporal, but to focus anew on the relations between the two. As my account in the preceding chapters discloses, temporality is often manifest in the spatial definition of everyday experiences and activities, whether this be agriculture or other economic tasks, ritual events like a jatre or wedding, life-cycle ceremonies, or even menstrual periods. Temporality is marked and played out in terms of events that punctuate the rhythms of one's life—an indexing that perhaps differs from the perception of time as dynamic and historical (see Bourdieu, 1990b). This is definitely not to reproduce the distinction between the historical and the experiential, but rather to suggest that the very modes in which we receive the categories of time and space have to be queried.

2. Analyzing domestic ceremonies elaborated in the *Grihya Sutras*, she finds a clear stress on the use of right and left sides in ritual, and suggests that ideas of auspiciousness and inauspiciousness are expressed in this idiom. Intersecting with these are the spatial co-ordinates of east–west–north–south. It could be pointed out that the above analysis is not enough to address the spatial question, as there are several other markers of space in addition to laterality and directionality. Some of these possibilities are explored in the following sections.

3. Recall also Note 1 above. Again, see the considerations which follow in the text.

4. Rosaldo (1974: 23) defines the domestic domain as 'those minimal institutions and modes of activity that are organized...around...mothers and their children'.

Eight

Postfix: A Sociology of the Body?

It should be obvious that certain worries about sociology/anthropology as also certain images of the imbrication of space with gender run through most of the book. It is also obvious, when approached schematically, that the chapters raise some pressing questions which they do not entirely answer. The ideas which occur here certainly need a rather more systematic framework, and I hope to be able to generate work in that direction over the next few years. It is this question of a framework—whose contours I discern, fairly conventionally, as a sociology of the body— that I would like to broach here by way of a supplement, a postfix. I do not think, however, that such a framework could have helpfully preceded these ideas; in fact, I think the framework will have to draw on these ideas.

Appropriating the Body

The place of the body within sociology particularly has been a rather ill-defined and ambiguous one. This could partly be due to the discipline's concentration on an abstract model of the human actor and an associated perception of culture as superorganic, both of which spell 'a denial of the somatic' (Jackson, 1983: 328) in sociology.[1] In the anthropological tradition, of course, the script has been fairly different. The most dominant trend there has been a representational or symbolic one, where the body is perceived as a sign or code signifying something else, be it cultural meanings or societal patterns. Suggestive as it might be to speak of the body as representing encoded social meanings—as an 'image of society' or even as a 'metaphor for society',[2]—the question remains as to whether these perspectives can acknowledge the materiality of bodies. When Douglas (1978: 87), for instance, suggests that 'in its role as an image of society, the body's main scope is to express the relation of the individual to the group', it is reduced to an instrument of communication. It is such a valorization of semiotic meaning that contributes, albeit indirectly, to an erasure of the actual body in analyses of social life. In other words, examining how bodies are formed or represented both by and within a culture is not enough to tell us about the spatiality of bodies, of living-*in*-bodies and living-*as*-bodies.

How one should begin speaking of such a lived 'materiality' of the body, at once spaced and spacing, is doubtless a troubling question. It perhaps needs to be clarified that reference here is not to the body 'as such'—which could steer us into the realm of philosophical discourse— but to the challenge of elaborating a more comprehensive socio-anthropo-logical approach to the body. This entails a scrutiny not merely of how the body is 'coded', but also demands an attentiveness to the practices which systematically constitute and animate the body. This would, in turn, take us beyond perspectives that either work with an idea of the body as 'given' or as somehow 'constructed'.

A substantial part of feminist anthropology has hitherto sought to deal with the problems posed by the body by choosing to align itself with the constructivist axis. This has meant taking a specific stand vis-à-vis the sex/gender distinction. More often than not, the trend has been to move away from a grounding in sexual difference, that is, in a biologically differentiated male and female, and towards an analysis of gender. The argument is that the grounding of 'women' as a biological category not only tends to essentialize gendered traits, but also fails to recognize the innumerable differences amongst women themselves, based on their caste, class or kinship positions. In keeping with this emphasis, studies have increasingly focused on the different ways in which femininity and masculinity are culturally constituted across diverse socio-cultural contexts. The outcome has been a steadily growing body of work which one may describe as a feminist symbolic anthropology (see, for instance, Ortner and Whitehead, 1981). In their preoccupation with the differential construction of gender and sexuality, these efforts seem to have deflected attention away from the female body itself to the layers of symbolic representation in which the body gets swathed.[3]

Though I am by no means claiming that this symbolic turn in anthro-pology is *the* most important genre of gender analysis, I refer to it largely because.it illustrates, in so many ways, the problems of persisting with a sex/gender distinction. The reluctance, within such studies, to speak about the female body other than as a symbol has led to a focus on femininity/

sexualization as an acquired trait, as an attribute that is socially inscribed, or as a form 'imposed on a sexed body'. And indeed, as critics of the sex/gender distinction (Gatens, 1992) have argued, the liberative dimension of gender in an earlier phase of scholarship seems now to be actually constricting analysis. It seems to me that delinking sexual difference from gender is, in effect, to formulate femininity/sexualization as an arbitrary form of behaviour and translates into a view of the body as irrelevant in comprehending its manifestations. In fact, taken to their logical extreme, words like 'inscribing' or 'imposing' assume the body as a kind of blank slate, as nothing apart from the cultural meanings constituting it. To be sure, the body is not quite such a passive receptacle but is the very medium through which meanings are produced. It is imperative, therefore, that we strive to re-establish the connections between the two principal schema defining and orienting female subjects in specific socio-historical contexts, namely, femininity/sexualization as an acquired attribute and femininity/sexualization as grounded in the female body.

Perhaps one way in which this restoration of both biology and ideology to the constitution of female identities can be achieved is to work with a notion of the female body as 'situated', embodied, in space, time, and culture. Along this axis—and very much in contrast to some of the formulations reviewed above—the sexed body is neither a mere symbol nor an inert biological foundation onto which gendered ideas are superscribed. Rather, such a situated body can be barely distinguished from the very techniques of the body that are retraceable through the bodily practices of women (be it with reference to the modes in which they work, walk, talk and dress, or even in relation to ritual routines). The body thus becomes the very medium through which femininity is spatialized and sexualization ordered. In many ways, this starting point, by blending the biological fact of womanhood with its ideological or cultural dimension, works to refuse superficial separations between sex/gender (or even body/mind).[4]

I must hasten to mention that such a move is not unrelated to developments in the 1980s and 1990s which have witnessed a phenomenal surge of interest in the body. Departing from earlier trends which either posited the body as a tangible, pre-discursive entity or as a purely representational medium, an increasing number of recent studies focus on the body-as-subject.[5] While the intellectual traditions they draw upon are indisputably diverse, they broadly concentrate upon the embodiment of subjects, bodily praxis and experientiality, rather than on bodily symbolism alone.

Nevertheless, it seems to me that the analytical implications of focusing on the body have not been uniform across these orientations. Falk (1994), for instance, prefaces his study of *The Consuming Body* with a distinction between the body and corporeality, the latter constituting bodiliness as the very *experientiality* of the body. As Falk observes, the focus on corporeality implies:

> *...an emphasis on the relation and dynamic role of the human body (and a distancing from the biological body conception) in the web of social and cultural liaisons; it concerns more specifically the effects of these liaisons on the experiential aspect of the body—the body as a sensory and sensual being* (Falk, 1994: 2).

While this standpoint represents one mode of appropriating the body question, a more basic consideration would be to address the larger context (or distinctive historical setting) that frames such studies on corporeality. The growing visibility of the body across disciplines is often seen as a modern (or postmodern?) symptom where questions of the self have become all-important, lending perspective to the idea that a person's sense of self is largely consolidated through the body. As Turner (1994: xii) summarizes, 'the project of the self in modern society is in fact the project of the body'. The truth of this observation is immediately apparent, whether we consider Foucault's (1978) analyses of sexuality or turn to feminist debates on gender identity. In thematizing the constitution of the self/subject, Foucault maintains that bodies are constituted within specific discourses of power or cultural regimes and, consequently, that there is nothing prior to these inscriptions. Yet, as Butler (1989) points out, to focus on the construction of the self or sexuality, even to speak of inscription on a surface, implicitly brings up the body as a pre-given, existentially available reality. She observes (and the passage is worth quoting at some length):

> *To claim that 'the body is culturally constructed' is, on the one hand, to assert that whatever meanings or attributes the body acquires are in fact culturally constituted and variable. But note that the very construction of the sentence confounds the meaning of 'construction' itself. Is the 'body' ontologically distinct from the process of construction it undergoes? If that is the case...(then), 'the body' would not be constructed, strictly considered, but would be the occasion, the site, or the condition of a process of construction only externally related to the body that is its object* (Butler, 1989: 601).

It will clearly not do then, to simply posit that complex relations obtain between the body and self or person. Indeed, what is being debated here concerns the very grounds of the spacing of bodies and persons. It has become increasingly apparent today that the sources and bases of personhood are neither solely physiological nor entirely a matter of consciousness; rather, they approximate to an idea of 'embodiment', that is to say, the active inhabiting of bodily and gendered spaces by distinctly situated subjects.

Embodiment

Whether it is in anthropology, feminism or any other discursive tradition, we clearly need to transcend the tendency to align exclusively with either naturalistic (read, biological) or constructivist arguments. As should be discernible, a focus on embodiment—on the active taking on and inhabiting of gendered spaces and identities—is one way of retaining the links between the two tendencies, although even here there is every possibility of a tilt towards any one direction. In the Indian context, for instance, a great deal of significant work regarding the construction of female sexuality continues to take place, addressing in particular the intersections of gender with caste, class, political events and socio-cultural practices (see Sangari and Vaid, 1989; Sundar Rajan, 1993; Tharu and Niranjana, 1994). While most of these elaborate on the intricate and complex moves through which communities construct female bodies, it is only very recently that attention has been paid to the 'lived body' in the everyday lives of women. Focusing on gender identities as both represented and lived, the essays in the volume edited by Thapan (1997) emblematize this recent shift in a theorizing geared towards women's experiences of their embodiment. While the female body definitely provides a common central core to these essays, the modes in which this space is negotiated varies considerably, moving from how communities and nations construct bodies to the lived body of everyday life, from regulated bodies to resisting bodies to inscribed bodies. This lends itself to varied notions of female embodiment, although, overall, they are underwritten by the same imperative of fusing the constructed and lived aspects of gender identities. What is also clear is that the preoccupation with embodiment is largely tied to the need to comprehend the working of gender inequality. In the words of Thapan (1997: 26): 'The emphasis, generally speaking, is on the nature of containment, regulation, manipulation, control and agency that women experience, both from external as well as inner sources, in relation to their embodiment.' The suggestion, avowedly, is that the 'woman's body is... central to understanding unequal gender relations' (Thapan 1997: 3).[6]

My own concern has been somewhat tangential to this emphasis, being geared towards highlighting an axis rarely addressed, namely, how female embodiment occurs through and in relation to space. Implicit to this thrust has been the necessity of rethinking the very meaning of construction itself, while lending oneself to an account of the 'materialization' of bodies. Again, Butler's (1990) notion of gender performativity could offer some perspectives on this score, indicating as it does the ways in which bodily norms are assumed by gendered subjects. Working against the grain of theories which understand gender identities as impositions from without, as inscriptions upon a passive body, Butler suggests that there is in fact an active taking on of identities and subject positions. But, as she clarifies in a subsequent work (Butler, 1993), this is not to be approached as presupposing an individual who deliberately chooses a particular gender role; instead, the subject who enacts gender, in being grounded within a materiality of the body, is always already constructed. What we have is not a polarity between discursive constructions on the one side and lived bodies on the other, but a situation where discourses live in and through the body. Rather than a pre-discursive body that subsequently adopts gender, what is indicated is that bodies shape themselves in the course of re-tracing gendered normative frames. As such, for Butler (1993: 2), performativity is a 'reiterative and citational practice' that reproduces the regulatory norms of sex and is not an enactment of choice. It ensures the materialization or legitimation of bodies that conform to a feminine ideal. And yet, in authorizing such bodies, Butler notes that the 'abjection' of other kinds of non-conforming bodies is also taking place. She argues that in 'assuming' a sex one is necessarily engaging with a 'heterosexual imperative' (Ibid.: 3) that offers certain kinds of sexed identifications and forecloses others; indeed, those who refuse such clear-cut identifications are not only excluded from the domain of subjects but also have no way of culturally articulating their difference. By suggesting that heterosexual normative codes work to constitute gendered identities and bodies, she is also concerned with addressing the question of how homosexuality becomes the abject 'other'.

The emphasis, however, comes to be slanted differently in the context issuing off Aaladapalya. Though there are clear-cut socio-spatial practices that are engaged in the regulation of female bodies, these operate primarily by 'legislating' between different orders of female sexuality. Thus, the image of the married, reproductive woman and the spaces occupied by her underscores a certain ideal femininity in relation to which the 'others'—girls, unmarried women, infertile women and widowed women—come to be marked. The overweaning importance of notions of shame and honour—which impinge on women's bodily comportment, speech and physical movement—draw on corresponding ideas of appropriate spaces for women. Put differently, female sexuality either acquires a legitimacy only within certain spaces, say marriage, and not outside, or in relation to certain moral codes that retrace a dominant normativity. It needs to be reiterated that this is not quite the same as the heterosexual matrix that Butler cites. Two complications could be noted here. First, the constitution of female bodies and identities via what has been described as the matrix of sexualization does not proceed solely in terms of drawing out its opposition to the male. This prompts one to ask whether Butler's insistence that the heterosexual imperative forms the major reference point for the construction of gendered bodies is not deflecting attention away from the other modes through which this construction simultaneously proceeds. The actual regulation of female bodies seems to be taking place beyond the broad strokes of male–female differences that the heterosexual matrix paints, that is, by legislating between different orders of femininity itself. This brings us to the second point: that the construction of female bodies also occurs through a delineation of the spaces through which 'regular' femininity ought to manifest itself. That is, in addition to a broad male–female distinction, women come to be defined in terms of perceptions of a 'proper' and 'improper' femininity as well. The female, or feminine, does not constitute an undifferentiated locus of sexualization; rather, there is a constant attempt to weigh and legislate between diverse subject positions within the feminine itself, all of which lends itself to the consolidation of an ideal female self, inhabiting socially approved spaces in modes that are culturally enjoined. As the discussions in the preceding chapters have shown, the axis of space has been central to the engendering of female bodies. It is not merely a marker of physical spaces, but also institutes a ground of sexualization through which women's activities, speech and bodily practices are mediated. In the course of negotiating the grid of olage–horage, women are not only constituting themselves as gendered selves, but are also retracing the contours of a dominant normativity.

All this allows us to suggest that there are socio–spatial undercurrents to performativity that render the negotiation of bodies and bodily identities as distinctly spatial and spatializing in their effects and embodiment. It seems to me that this is an axis relatively undertheorized in Butler and would require us to ask how female subjects orchestrate their lives as women through the body, in addition to querying how ideas of femininity are internalized. Such a widening of the axis of theorization would involve several levels of investigation—such as how women inhabit the body, what the socio-cultural meanings invested in the female body are, how women's bodies occupy and orient themselves in space, and so on— some aspects of which have been disclosed in the foregoing chapters. To work more systematically with an idea of embodiment, with a notion of the 'body as situated', however, would enable us to explore these questions at length, since the body now comes to be appropriated less as 'subject' and more pertinently as a *'situated* subject', a 'lived body'.[7]

Body(–)Space

While these formulations are suggestive of an analysis issuing from an interface between the material body, the representational body and the modes in which women activate these constitutive conditions, the alternative view of femininity and embodiment that this can offer is grounded in a very specific claim—the claim for a *spatializing* of bodily discourses and practices. Such a tracing of the spatial dimension calls attention to the contexts in which subjects live their lives, the arenas, events or qualities that mark bodies as female (or male), as well as how the body itself condenses location in a cultural space. Spatializing this discourse of the female body would also make it possible to disaggregate the analysis of gender without necessarily falling back upon the binarism of the body as either culturally constituted or as prior to its cultural inscription.

Such a framework would necessarily draw on cultural discourses as they impinge on the body and yet seek to tease out 'habitual patterns of body use'. Turner's (1994: x) bald assertion that inspite of everything 'we still lack a general theory of the body in society and of society in the body' must however be queried. A sociology of the body certainly needs the benefits of theory, but I am convinced that what it does not need is a 'general theory'. There cannot be any neat or self-contained theory of what the body is, though there can be a matrix of appraisal for contending with the body which issues off personal experiences and social institutions. Seeking such a matrix has been central to this book, in addition to other related preoccupations. The abstract and schematic conceptions of space and gender routinely deployed have had to be challenged, and this has meant drawing on lived discourses as they impinge on the body while teasing out patterns of body use and bodily inscription. It is these patterns, at once spatial and spatializing, that combine the societal with the individual planes, where bodily articulations merge into ideas about the socio-spatial world. My intention, of course, has not been to project the body as some kind of an entity, but to work towards developing a perspective on it as a sort of body space: to suggest that bodiliness or how one experiences the body (and its identity) is mediated through a socio-spatial matrix, which in the vernacular is rendered as olage–horage. It is this matrix that anchors the everyday, giving rise to certain bodily and mental dispositions which, while being informed by shared social habits, are 'collectively orchestrated without being the product of the orchestrating action of a conductor' (Bourdieu, 1977: 72). Mapping forms of and variations within this body–space would not only make it possible to throw light on the nature and forms of embodiment within communities, but would also be expressive of how embodied subjects come to inhabit and constitute social and physical spaces. Equally important would be an exploration of variations in this grid over time and the manner in which this affects the gendering of selves. To assess these claims and to compare them across a variety of settings remains a central and pressing demand on scholarship.

Notes

1. Goffman's (1971) work on the presentation of self in everyday life is perhaps an exception. An understanding of the body is crucial to his concept of identity, but is hardly worked through. Bourdieu (1977) is the other major thinker who is attentive to the body.
2. See, for instance, the analysis of laterality in Hertz (1960) and the prognosis offered of caste and the female body by Das (1988a). Douglas (1970, 1986) is also illustrative.
3. What is being indicated here is not the existence of the female body as some pre-discursive, already–given foundation, but the need to reclaim it as a sort of materiality, that is, as a force actively negotiating the bounds of normative behaviour.
4. At issue here is also the concept of the 'individual' one is working with. Instead of deploying an abstract formulation of the individual human being, what is needed is an attentiveness to the bodily basis of personhood and its links with culture. Gender identities, then, are best understood as an embodiment, connoting the confluence of material bodies and cultural practices. For the body/mind dualism, see, Held (1990).
5. To cite some among many: Bordo (1990), Butler (1990, 1993), Bynum (1995), Frank (1990), Grosz (1994), Martin (1987) and Nancy (1994).
6. Note also that this is a point most explicitly raised by feminist activists in tackling issues like women's health, sexual violence, abuse, and so on.
7. What is the idea of the 'subject' that is entailed here? Notions of the subject, subjecthood, subjection, subjectivity and so on have been the locus of wide-ranging debate, involving philosophers, political and social scientists, feminists and others.

While it would be impossible to do justice to the various strands of thought here, one could underline some central and recurrent issues: (*a*) the conferral or assumption of subjecthood, seen as always taking place within a specific circuit of power relations; (*b*) subjecthood as involving subjection/submission, usually 'as willing obedience, coming from inside' (Balibar, 1994: 9); and (*c*) subjecthood as creating the space for subjectivity, for an exercise of the liberty and creativity of a human being. To speak of the 'body as subject', even more of a *situated* subject, is to draw on all these resonances of the term—as constituted, within a given matrix of discourses and regulatory practices, and as constituting, that is, as determining the very ground of these practices and discourses. Underwriting these activities of 'constituting' and 'being–constituted' are of course distinctive spatial coordinates.

Appendix

Aaladapalya: A Socio-material Profile

In the past decade or so, the study of gender questions has increasingly drawn attention to the fact that the household cannot be treated as an indivisible unit of analysis. This implies the fairly obvious point that questions about the distribution of resources within the household matter independently of their distribution between households. But, more subtly, it can imply that the ways of influencing the distribution of resources between households have a different impact on what happens within. In serving up this socio-material profile of Aaladapalya, therefore, my intention is not to *complete* the description of the field; rather, it is to provide a substantive buoy—anchoring as well as keeping afloat—the ethnographic traversal of the foregoing pages.

Physical Layout

Aaladapalya, as my chapters have disclosed, is a cluster of two villages and four hamlets, all located about 30 kilometres west of Bangalore city. Spread over a radius of 2 to 3 kilometres, the settlements are fairly old and essentially Hindu. They are multi-caste, although occupations are not very diverse. The mainstay is agriculture, with land being the pivot of their world-view.

The two main villages (Doddaaluru and Hulaguppe) consist primarily of a core cluster of households encircled by fields. In contrast, the hamlets (Arasapura, Chikkuru, Shivapura and Budkepalya) are not only much smaller, but also dispersed, comprising of households that have physically separated from the core settlement in order to live close to their land. Within each village, residential patterns both echo and shape societal organization, yielding information about the intensity of interactions within the family and the nature and frequency of relations with neighbouring households and castes. More obviously, they are also a reflection of rules of caste precedence and dominance. For instance, while the central part of the village is occupied by castes comparatively higher in the caste hierarchy and scheme of economic precedence (like the Lingayats and Vokkaligas), the lower-castes (especially the Dalits) inhabit the fringes of the settlement.

Doddaaluru is best described as a linear village cluster. Houses—belonging mainly to the middle- and upper-castes—are laid out on either side of the main road, behind which there are parallel rows of more houses. The *madiga keri* (Dalit colony) is separate from, and at right angles to, the rest of the habitat. The major castes in this village are the Madigas, Vokkaligas, Lingayats and Kurubas. Ganigas (oilpressers) and Acharis (smiths) comprise a few households. A majority of the people depend either wholly or in part on agriculture, that is, as independent farmers or as wage labourers.

Arasapura is a tiny hamlet about 1 kilometres to the north of Doddaaluru. It consists of ten households, five of which belong to the Vokkaliga caste, while the remain-ing are Dalit families. Most of them moved to this settlement about a generation or two ago in order to be adjacent to their lands.

Hulaguppe is one of the largest villages in the region, with 200 households and a population of around 1,200. It consists of a core settlement and some out-lying hamlets (Shivapura, Chikkuru and Budkepalya), all of which constitute Hulaguppe, administratively defined. The core settlement has 135 households and a range of castes like Lingayats, Dalits (both Madigas and Holeyas), Vokka-ligas, Brahmins, Bajentris, Kurubas, Madivalas, and so on. The clearest and most visible divide, however, is between the Dalits (who form nearly half the total number of households) and the other castes. Cultivation of land is the predominant activity here. Some households (like the washermen) pursue their traditional occupations, but tend to supplement their earnings through agricultural labour. A segment of the male population holds salaried jobs in the city.

Today, a main road touches the hem of the village and a regular bus service plies between Bangalore and Hulaguppe. It has a post office as well as a primary health centre catering to the needs of several surrounding villages and hamlets. A government-run primary and middle school for boys and girls is located here, while a high school is run under private management. An agricultural co-operative society also exists, which provides loans for purchase of seeds and fertilizers to its members and shareholders; marginal farmers can also avail of subsidies. Increasingly, Aaladapalya has been exposed to a number of developmental programmes as well due to the activities of non-governmental organizations and social work agencies.

While the hamlets of Shivapura and Chikkuru are offshoots of Hulaguppe, Budkepalya represents a marked contrast. Unlike the other hamlets and villages, this is a recent settlement, barely 10 years old. It emerged as a result of prolonged efforts by social work organizations to 'settle' the nomadic Budabudike tribals by providing them with houses and a means of livelihood in the form of land. Having been a nomadic people for many generations, these people do not have any specific regional concentration. Clusters of families can be found in Bangalore, Kanakapura, Mandya, Koppa, Dharwad, Hubli and other places in Karnataka. Some families reside adjacent to the Banyan tree in Aaladapalya. In fact, most of the twenty-odd families at Budkepalya are splinters of the core households residing near the Banyan tree, who moved when offered facilities at the new settlement.

Caste

In Aaladapalya as a whole, the major caste groups are the Lingayats and Dalits, followed by a sizeable number of Vokkaliga households. In addition to these, there are a few service castes. A detailed break-up of caste groups in terms of households is:

Lingayat	88
Dalit	83
Vokkaliga	40
Kuruba	22
Budabudike	21
Others	25
Total	279

The Budabudike are not strictly a caste and are classified as a tribe (though for all social purposes they are treated as a lower-caste). The category 'Others' includes six Brahmin households, four each of Acharis and Bajentris, three Banajiga households, two each of Ganiga, Bovi and Idiga, and one each of Madivala and Padmashali. It is evident that a bulk of the households—about 61 per cent—are either Lingayat or Dalit, with each having approximately the same numerical preponderance. However, this trend is not the same in individual villages, since caste groups are differentially distributed across the village cluster. The caste-wise distribution of households in terms of the various settlements comprising Aaladapalya is given in Table A1.1

Table A1.1: Distribution of Households

Settlement/ Caste	Doddaaluru	Hulaguppe	Arasapura	Chikkuru	Shivapura	Budkepalya
Lingayat	19	55	–	8	5	1
Dalit	19	49	5	–	10	–
Vokkaliga	16	9	5	–	9	1
Kuruba	21	1	–	–	–	–
Budabudike	–	–	–	–	–	21
Others	6	11	–	–	3	5
Total	81	125	10	8	27	28

Economic Standing

Landownership remains the most significant marker of socio-economic standing. The extent of landownership, as reported by respondents, is also indicative of subjective perceptions of status. A classification may be attempted as under:

Landless households	62
1 acre and below	50
Above 1 acre–up to 2 acres	58
Above 2 acres–up to 4 acres	61
Above 4 acres–up to 10 acres	40
Above 10 acres–up to 20 acres	8

A clear correlation obtains between caste and landownership (see Table A1.2), with 45 per cent of the landless households being Dalit, while a majority of those owning between 5 and 20 acres being Lingayats (followed by Vokkaligas). Even as 22 per cent of the landless are Lingayats, they nevertheless command a high position due to salaried jobs in nearby towns and the city.

Table A1.2: Caste and Landownership

Caste/Size	Lingayat	Dalit	Vokkaliga	Kuruba	Budabudike	Others
Landless	14	28	6	1	5	8
1 acre and below	8	27	–	4	5	6
Above 1 acre– up to 2 acres	12	18	8	6	7	7
Above 2 acres– up to 4 acres	29	8	10	8	3	3
Above 4 acres– up to 10 acres	21	2	13	2	1	1
Above 10 acres– up to 20 acres	4	–	3	1	–	–

Family and Kinship Structure

The size of households in Aaladapalya is as follows:

Less than two members	13
Between 3–5 members	155
Between 6–8 members	89
Above 9 members	22

The fact that over 50 per cent of the households have between three to five members is itself indicative of the non-joint nature of the family structure. This can perhaps serve as a backdrop for understanding and explaining the divergences in women's roles and power within the household. The main defining criteria of different family types have usually been identified as commensality and residence. Yet, specifying the relatives who constitute the family is equally significant in defining the type of household. Our categorization of household types which follows draws selectively from Kolenda's (1987) classificatory scheme.

Single person household	3
Nuclear family	170
Supplemented nuclear family	50
Sub-nuclear family and supplemented sub-nuclear family	12
Collateral joint family and supplemented collateral joint family	6
Linear joint family	33
Lineal-collateral joint family and supplemented lineal-collateral joint family	5

It is again evident that a large majority of the families are nuclear. Nuclear families supplemented by unmarried or widowed relatives of parents form another significant block, and these two household types together account for nearly 79 per cent of the field. Lineal joint families, where parents and married children live together, comprise about 12 per cent of the total households. The tendency toward nuclear families, however, must be seen in the context of the prevailing economic structure, where there is extensive impoverishment and the percentage of landless and marginal farmers is very high.

Women's Work

Analyzing the different kinds of work done by women not only throws light on their negotiation of spaces inside and outside the household, but also reveals significant differences in relation to caste and economic standing. The work sphere of upper-caste women, for instance, is largely confined to the household (nearly 44 per cent of Lingayat women) and/or to the family land. In contrast, women from lower castes and classes engage in a multiplicity of jobs to supplement their incomes. These range from agricultural tasks on family land and on others' land (for a wage) to other kinds of tasks (*coolie kelasa*) like road construction, at a brick kiln, and so on. Yet, the range of jobs available to them, as also their choice of jobs, are underwritten by a number of considerations. As we argued in Chapter Three, questions of morality are often brought to bear on women undertaking work outside the household. In addition to this, the extent of land owned by the household also affects the nature of women's work. A substantial number of women who do not undertake wage labour but work only on family land (nearly 35 per cent) are from small and middle farmer households. At higher levels of landownership, women commit themselves largely to housework. In stark contrast, a majority of the women from landless households rely on wage labour of different sorts. Attending to these differences make it possible to see how women's negotiation of spaces in the course of their work is overlaid with both moral and material significances. What is more, they lend further perspective to the point that it makes no sense to focus on the household as an indivisible unit of analysis and to treat what happens within the household as the province of a different sphere of action.

Bibliography

Agnew, J. 1993. 'Representing space: Space, scale and culture in social science', in J. Duncan and D. Ley, eds, *Place/culture/ representation*. Routledge: London, pp. 251–71.

Appadurai, A. 1988a. 'Introduction: Place and voice in anthropological theory'. *Cultural Anthropology*, 3(1): 16–20.

———. 1988b. 'Putting hierarchy in its place'. *Cultural Anthropology*, 3(1): 36–49.

Ardener, E. 1975. 'Belief and the problem of women and the "Problem" revisited', in S. Ardener, ed., *Perceiving women*. London: Malaby Press, pp. 1–27.

Ardener, S., ed. 1975. *Perceiving women*. London: Malaby Press.

———. 1981. *Women and space*. London: Croom Helm.

Asad, T. 1973. 'Introduction', in T. Asad, ed., *Anthropology and the colonial encounter*. London: Ithaca Press, pp. 9–19.

Bailey, F.G. 1957. *Caste and the economic frontier*. Manchester: Manchester University Press.

———. 1959. 'For a sociology of India?'. *Contributions to Indian Sociology*, 3: 88–101.

Balibar, E. 1994. 'Subjection and subjectivation', in J. Copjec, ed., *Supposing the subject*. London: Verso, pp. 1–15.

Beteille, A. 1965. *Caste, class and power*. Berkeley: University of California Press.

———. 1992. 'Individualism and equality', in A. Beteille, ed., *Society and politics in India: Essays in a comparative perspective*. Delhi: Oxford University Press, pp. 215–49.

Beteille, A. and Madan T.N., eds. 1975. *Encounter and experience: Personal accounts of fieldwork*. Delhi: Vikas Publishing House.

Bordo, S. 1990. 'Feminism, postmodernism and gender-scepticism', in L. Nicholson, ed., *Feminism/postmodernism*. New York: Routledge.

Bourdieu, P. 1977. *Outline of a theory of practice*. Cambridge: Cambridge University Press.

———. 1990a. *The logic of practice*. Cambridge: Polity Press.

———. 1990b. 'Time perspectives of the Kabyle', in J. Hassard, ed., *The sociology of time*. Basingstoke, Hampshire: Macmillan, pp. 219–37.

Busby, C. 1997. 'Permeable and partible persons: A comparative analysis of gender and body in South India and Melanesia'. *Journal of the Royal Anthropological Institute*, 3(2): 261–77.

Butler, J. 1989. 'Foucault and the paradox of bodily inscriptions'. *The Journal of Philosophy*, 86(11): 601–07.

———. 1990. *Gender trouble: Feminism and the subversion of identity*. New York: Routledge.

Butler, J. 1993. *Bodies that matter.* New York: Routledge.

Bynum, C. 1995. 'Why all the fuss about the body? A medievalist's perspective'. *Critical Inquiry*, 22(1): 1–33.

Carman J.B. and Marglin, F.A., eds. 1985. *Purity and auspiciousness in Indian society.* Leiden: E.J. Brill.

Chatterjee, P. 1989. 'On the nationalist resolution of the women's question', in K. Sangari and S. Vaid, eds, *Recasting women.* New Delhi: Kali for Women, pp. 233–53.

Clifford, J. 1986. 'Introduction: Partial truths', in J. Clifford and G.E. Marcus, eds, *Writing culture: The politics of ethnography.* Berkeley: University of California Press, pp. 1–26.

———. 1997. 'Spatial practices: Fieldwork, travel and the discipline of anthropology', in A. Gupta and J. Ferguson, eds, *Anthropological locations: Boundaries and grounds of a field science.* Berkeley: University of California Press, pp. 185–222.

Clifford, J. and Marcus, G.E., eds. 1986. *Writing culture: The politics of ethnography.* Berkeley: University of California Press.

Connell, R.W. 1987. *Gender and power.* Stanford: Stanford University Press.

Daniel, E.V. 1987. *Fluid signs: Being a person the Tamil way.* Berkeley: University of California Press.

Das, V. 1982. *Structure and cognition.* Delhi: Oxford University Press.

———. 1988a. 'Feminity and the orientation to the body', in K. Chanana, ed., *Socialization, education and women: Explorations in gender identity.* Delhi: Oxford University Press, pp. 193–207.

———. 1988b. 'Shakti versus sati: A reading of the Santoshi Ma cult'. *Manushi*, 49: 26–30.

———. 1996. *Critical events: An anthropological perspective on contemporary events.* Delhi: Oxford University Press.

Desai, A.R., ed. 1969. *Rural sociology in India.* Mumbai: Popular Prakashan.

Dirks, N.B. 1987. *The hollow crown: Ethnohistory of an Indian kingdom.* New York: Cambridge University Press.

Douglas, M. 1970. *Natural symbols: Explorations in cosmology.* London: Barrie and Jenkins.

———. 1978. *Implicit meanings: Essays in anthropology.* London: Routledge and Kegan Paul.

———. 1986. *Purity and danger: An analysis of concepts of pollution and taboo.* Harmondsworth: Penguin.

Dube, L. 1986. 'Seed and earth: The symbolism of biological reproduction and sexual relations of production', in L. Dube, E. Leacock and S. Ardener, eds, *Visibility and power: Essays on women in society and development.* Delhi: Oxford University Press, pp. 22–53.

———. 1996. 'Caste and women', in M.N. Srinivas, ed., *Caste: Its twentieth century avatar.* New Delhi: Viking, pp. 1–27.

Dube, S.C. 1967. *Indian village.* Mumbai: Allied Publishers Pvt. Ltd.

———. 1969. 'The study of Indian village communities', in A.R. Desai, ed., *Rural sociology in India.* Mumbai: Popular Prakashan, pp. 790–95.

Dumont, L. 1966. 'The village community from Munro to Maine'. *Contributions to Indian Sociology*, 9: 67–89.

———. 1970. 'A structural definition of a folk deity of Tamilnad: Aiyanar, the Lord', in L. Dumont, ed., *Religion, politics and history in India: Collected papers in Indian Sociology.* The Hague: Mouton, pp. 20–32.

Dumont, L. 1980. *Homo hierarchicus: The caste system and its implications*. Chicago: University of Chicago Press.

———. 1986. *A South Indian subcaste: Social organization and religion of the Pramalai Kallar*. Delhi: Oxford University Press.

Dumont, L. and Pocock, D.F. 1957. 'Village studies'. *Contributions to Indian Sociology*, 1: 23–42.

———. 1959. 'On the different aspects or levels in Hinduism'. *Contributions to Indian Sociology*, 3: 31–44.

Duncan, N., ed. 1996. *Bodyspace: Destabilizing geographies of gender and sexuality*. London: Routledge.

Epstein, T.S. 1979. *Economic development and social change in South India*. London: J. K. Publishers.

Erndl, K.M. 1993. *Victory to the Mother: The Hindu goddess of northwest India in myth, ritual and symbol*. New York: Oxford University Press.

Falk, P. 1994. *The consuming body*. London: Sage.

Foucault, M. 1978. *History of sexuality, Vol. 1*. New York: Vintage.

———. 1986. *Power/knowledge*. Sussex: Harvester Press.

Frank, A.W. 1990. 'Bringing bodies back in: A decade review'. *Theory, Culture and Society*, 7(1): 131–62.

Fruzzetti, L. and Ostor, A. 1976. 'Seed and earth: A cultural analysis of kinship in a Bengali town'. *Contributions to Indian Sociology* (n.s.), 10(1): 97–132.

Fuller, C.J. 1988. 'The Hindu pantheon and the legitimation of hierarchy'. *Man* (n.s.), 23(1): 19–39.

———. 1992. *The camphor flame: Popular Hinduism and society in India*. New Delhi: Viking, Penguin India.

Ganesh, K. 1989. 'Seclusion of women and the structure of caste', in M. Krishnaraj and K. Chanana, eds, *Gender and the household domain: Social and cultural dimensions*. New Delhi: Sage, pp. 75–95.

———. 1990. 'Mother who is not a mother: In search of the great Indian goddess'. *Economic and Political Weekly*, 25 (42 & 43): WS 58–64.

———. 1993. *Boundary walls: Caste and women in a Tamil community*. Delhi: Hindustan Publishing House.

Gatens, M. 1992. 'Power, bodies and difference', in M. Barrett and A. Phillips, eds, *Destabilizing theory: Contemporary feminist debates*. Stanford: Stanford University Press, pp. 120–37.

Gatwood, L.E. 1985. *Devi and the spouse goddess: Women, sexuality and marriage in India*. New Delhi: Manohar.

Giddens, A. 1982. *Sociology: A brief but critical introduction*. London: Macmillan.

Goffman, E. 1971. *The presentation of self in everyday life*. Harmondsworth: Pelican.

Grosz, E. 1994. *Volatile bodies: Toward a corporeal feminism*. Bloomington: Indiana University Press.

Gupta, A. and Ferguson, J., eds. 1997a. *Anthropological locations: Boundaries and grounds of a field science*. Berkeley: University of California Press.

———. 1997b. *Culture, power, place: Explorations in critical anthropology*. Durham, N.C.: Duke University Press.

Hanchett, S. 1988. *Coloured rice: Symbolic structure in Hindu family festivals*. Delhi: Hindustan.

Hegde, S. and Niranjana, S. 1994. 'Of the religious and the (non)-feminine: Open questions'. *Contributions to Indian Sociology* (n.s.), 28(1): 107–22.

Held, V. 1990. 'Feminist transformations of moral theory'. *Philosophy and Phenomeno-logical Research*, 50: 321–44.

Hertz, R. 1960. *Death and the right hand*. London: Cohen and West.

Hillier, B. and Hanson, J. 1984. *The social logic of space*. Cambridge: Cambridge University Press.

Jackson, M. 1983. 'Knowledge of the body'. *Man* (n.s.), 18(2): 327–45.

John, M.E. 1996. *Discrepant dislocations: Feminism, theory and postcolonial histories*. Delhi: Oxford University Press.

Kapadia, K. 1996. *Siva and her sisters: Gender, caste and class in rural South India*. Delhi: Oxford University Press.

Karlekar, M. 1995. 'Search for women's voices: Reflections on fieldwork, 1968–93'. *Economic and Political Weekly*, 30(17): WS 30–37.

Khare, R.S. 1976. *The Hindu hearth and home*. New Delhi: Vikas.

———. 1983. 'From *kanya* to *mata*: Aspects of the cultural language of kinship in Northern India', in A. Ostor, L. Fruzzetti and S. Barnett, eds, *Concepts of person: Kinship, caste and marriage in India*. Delhi: Oxford University Press, pp. 143–71.

Kolenda, P. 1987. *Regional differences in family structure in India*. Jaipur: Rawat Publications.

Kondor, V. 1986. 'Images of the fierce goddess and portrayal of Hindu women'. *Contributions to Indian Sociology* (n.s.), 20(2): 173–97.

Kumar, Krishan. 1978. *Prophecy and progress: The sociology of industrial and post-industrial society*. Harmondsworth: Penguin.

Kumar, N. 1992. *Friends, brothers and informants: Fieldwork memoirs of Banaras*. Berkeley: University of California Press.

Lefebvre, H. 1991. *The production of place*. Oxford: Basil Blackwell.

MacCormack, C. and Strathern, M., eds. 1980. *Nature, culture and gender*. Cambridge: Cambridge University Press.

Madan, T.N. 1965. Family and kinship—A study of the Pandits of Kashmir. Bombay: Asia Publishing House.

———. 1985. 'Concerning the categories *subha* and *suddha* in Hindu culture: An exploratory essay', in J.B. Carman and F.A. Marglin, eds, *Purity and auspiciousness in Indian society*. Leiden: E.J. Brill, pp. 11–29.

———. 1987. *Non-renunciation: Themes and interpretations of Hindu culture*. Delhi: Oxford University Press.

Malamound, C. 1981. 'On the rhetoric and semantics of purusartha'. *Contributions to Indian Sociology* (n.s.), 15(1&2): 33–54.

Manganaro, M., ed. 1990. *Modernist anthropology: From fieldwork to text*. Princeton, N.J.: Princeton University Press.

Marcus, G.E and Fischer, M.M.J. 1986. *Anthropology as cultural critique: An experimental moment in the human sciences*. Chicago: University of Chicago Press.

Marglin, F.A. 1985a. 'Types of oppositions in Hindu culture', in J.B. Carman and F.A. Marglin, eds, *Purity and auspiciousness in Indian society*. Leiden: E.J. Brill, pp. 65–83.

———. 1985b. *Wives of the god king*. Delhi: Oxford University Press.

Marriott, M., ed. 1961. *Village India*. Mumbai: Asia Publishing House.

———. 1990. *India through Hindu categories*. New Delhi: Sage.

Martin, E. 1987. *The woman in the body: A cultural analysis of reproduction*. Milton Keynes: Open University Press.

Massey, D. 1992. 'Politics and space/time'. *New Left Review*, 196: 65–84.

Mazumdar, V. 1985. 'Emergence of women's question in India and the role of women's studies'. Occasional Paper 7. New Delhi: Centre for Women's Development Studies.

Mazumdar, V. and Sharma, K. 1979. 'Women's studies: New perceptions and the challenges'. *Economic and Political Weekly*, 14(3): 113–20.

Meijer, I.C. and Prins, B. 1998. 'How bodies come to matter: An interview with Judith Butler'. *Signs: Journal of Women in Culture and Society*, 23(2): 275–86.

Misri, U. 1985. 'Child and childhood: A conceptual construction'. *Contributions to Indian Sociology* (n.s.), 19(1): 115–31.

Moore, H. 1988. *Feminism and anthropology*. Cambridge: Polity Press.

Nancy, J-L. 1994. 'Corpus'. in J.F. MacCannell and L. Zakarin, eds, *Thinking bodies*. California: Stanford University Press, pp. 17–31.

Narayan, K. 1993. 'How native is a "native" anthropologist? *American Anthropologist*, 95(3): 671–86.

Niranjana, S. 1991. 'Conceptualizing the Indian village: An overview of the village studies tradition'. *The Indian Journal of Social Science*, 4(3): 371–86.

———. 1992. 'Discerning women: Variations on the theme of gender—a review'. *The Indian Journal of Social Science*, 5(4): 393–412.

———. 1994. 'On gender and difference: Towards a re-articulation'. *Social Scientist*, 22(7–8): 28–41.

O'Hanlon, R. 1991. 'Issues of widowhood: Gender and resistance in colonial western India', in D. Haynes and G. Prakash, eds, *Contesting power: Resistance and everyday social relations in South Asia*. Berkeley: University of California Press, pp. 62–108.

Omvedt, G. 1980. *We will smash this prison: Indian women in struggle*. London: Zed Press.

Ortner, S. 1974. 'Is female to male as nature is to culture?', in M.Z. Rosaldo and L. Lamphere, eds, *Woman, culture and society*. Stanford: Stanford University Press, pp. 67–88.

———. 1984. 'Theory and anthropology since the sixties. *Comparative Studies in Society and History*, 26(1): 126–66.

Ortner, S. and Whitehead, H., eds. 1981. *Sexual meanings*. Cambridge: Cambridge University Press.

Raheja, G.G. 1988. *The poison in the gift: Ritual, prestation and the dominant caste in a North Indian village*. Chicago: University of Chicago Press.

Raheja, G.G. and Gold, A.G. 1996. *Listen to the heron's words*. Delhi: Oxford University Press.

Ram, K. 1992. *Mukkuvar women: Gender, hegemony and capitalist transformation in a South Indian fishing community*. New Delhi: Kali for Women.

———. 1999. 'Review of M. Thapan (ed.) *Embodiment: Essays on gender and identity*'. *Contributions to Indian Sociology* (n.s.), 33 (1&2): 450–52.

Rapp, R., ed. 1975. *Toward an anthropology of women*. New York: Monthly Review Press.

Rodman, M.C. 1992. 'Empowering place: Multilocality and multivocality'. *American Anthropologist*, 94(3): 640–56.

Rosaldo, M.Z. 1974. 'Woman, culture and society: A theoretical overview', in M.Z. Rosaldo and L. Lamphere, eds, *Woman, culture and society*. Stanford: Stanford University Press, pp. 17–42.

Rosaldo, M.Z. and Lamphere, L., eds. 1974. *Woman, culture and society*. Stanford: Stanford University Press.

Rose, G. 1996. 'As if the mirrors had bled', in N. Duncan, ed., *Bodyspace: Destabilizing geographies of gender and sexuality*. Routledge, pp. 56–74.

Said, E. 1985. *Orientalism*. London: Penguin.

Sanday, P. 1974. Female status in the public domain', in M.Z. Rosaldo and L. Lamphere, eds, *Woman, culture and society*. Stanford: Stanford University Press, pp. 189–206.

Sangari, K. 1993. 'Consent, agency and rhetorics of incitement'. *Economic and Political Weekly*, 28(18): 867–82.

Sangari, K. and Vaid, S., eds. 1989. *Recasting women: Essays in colonial history*. New Delhi: Kali for Women.

Scott, J. 1986. 'Gender: A useful category for historical analysis'. *American Historical Review*, 91(5): 1053–75.

Smith, N. and Katz, C. 1993. 'Grounding metaphor: Towards a spatialized politics', in M. Keith and S. Pile, eds, *Place and the politics of identity*. London: Routledge, pp. 67–83.

Soja, E.W. 1989. *Postmodern geographies: The reassertion of space in critical social theory*. London: Verso.

Srinivas, M.N. ed. 1955. *India's villages*. Calcutta: Government of West Bengal Publication.

———., 1962. *Religion and society among the Coorgs of Southern India*. Delhi: Oxford University Press.

———. 1966. 'Some thoughts on the study of one's own society', in M.N. Srinivas, *Social change in modern India*. Mumbai: Allied Publishers.

———. 1976. *The remembered village*. Delhi: Oxford University Press.

———. 1978. 'Varna and caste', in M.N. Srinivas, ed., *Caste in modern India and other essays*. Mumbai: Media Promoters and Publishers, pp. 63–69.

———. 1987. *The dominant caste and other essays*. Delhi: Oxford University Press.

———., ed. 1996a. *Caste: Its twentieth century avatar*. New Delhi: Viking.

———. 1996b. *Village, caste, gender and method: Essays in Indian social anthropology*. Delhi: Oxford University Press.

Srinivas, M.N., Shah, A.M. and Ramaswamy, E.A., eds. 1979. *The fieldworker and the field*. Delhi: Oxford University Press.

Strathern, M. 1981. 'Culture in a netbag: The manufacture of a subdiscipline in anthropology'. *Man* (n.s.), 16(4): 665–88.

———. 1987a. 'An awkward relationship: The case of feminism and anthropology'. *Signs*, 12(2): 276–92.

———. 1987b. 'The limits of auto-anthropology', in A. Jackson, ed., *Anthropology at home*. London: Tavistock, pp. 16–37.

Sundar Rajan, R. 1993. *Real and imagined women: Gender, culture and postcolonialism*. London: Routledge.

Tapper, B. 1979. 'Widows and goddesses: Female roles in deity symbolism in a South Indian village'. *Contributions to Indian Sociology* (n.s.), 13(1): 1–31.

Thapan, M. 1995. 'Gender, body and everyday life'. *Social Scientist*, 23(7–9): 32–58.

Thapan, M., ed. 1997. *Embodiment: Essays on gender and identity*. Delhi: Oxford University Press.

———. 1998. *Anthropological journeys: Reflections on fieldwork*. New Delhi: Orient Longman.

Tharu, S. and Niranjana, T. 1994. 'Problems for a contemporary theory of gender'. *Social Scientist*, 22(3–4): 93–117.

Thorner, D. 1966. 'Marx on India and the Asiatic mode of production'. *Contributions to Indian Sociology*, 9: 33–66.

Turner, B.S. 1984. *The body and society: Explorations in social theory.* Oxford: Basil Blackwell.

———. 1994. 'Preface', in P. Falk, ed., *The consuming body.* London: Sage, pp. vii–xvii.

Visweswaran, K. 1996. *Fictions of feminist ethnography.* Delhi: Oxford University Press.

Winslow, D. 1980. 'Rituals of first menstruation in Sri Lanka'. *Man* (n.s.), 15(4): 603–25.

Worsley, P. 1984. *The three worlds: Culture and world development.* London: Weidenfeld and Nicolson.

Index